THE PREVENTION OF OIL POLLUTION

THE PREVENTION OF OIL POLLUTION

Edited by
J. Wardley-Smith

Formerly Technical Manager, The International Tanker Owners' Pollution Federation

Graham & Trotman Limited

First published in 1979 by

Graham and Trotman Limited
Bond Street House
14 Clifford Street
London W1X 1RD

© Graham and Trotman Limited, 1979
Softcover reprint of the hardcover 1st edition 1979

ISBN-13: 978-94-011-7349-0   e-ISBN-13: 978-94-011-7347-6
DOI: 10.1007/978-94-011-7347-6

This publication is protected by international copyright law. All rights reserved. No part of this publication may be reproduced, stored in a retrieval system, or transmitted in any form or by any means, electronic, mechanical, photo-copying, recording or otherwise, without the prior permission of the publishers.

Contents

List of figures		vii
List of plates		ix
List of tables		x

Chapter 1	Sources of pollution J. Wardley-Smith		1
Chapter 2	Effect of oil spills on land and water A. Nelson-Smith		17
Chapter 3	Air pollution from the use of petroleum S. R. Craxford		35
Chapter 4	Pollution occurring during exploration and production on land and sea D. R. Blaikley		61
Chapter 5	Accidental spills from tankers and other vessels W. O. Gray		79
Chapter 6	Operational pollution from tankers and other vessels R. Maybourn		115
Chapter 7	Port operations: Spills during loading, discharging and bunkering R. S. Hawkins		133
Chapter 8	The pipelines contribution E. M. King		153

Chapter 9	Spills during transport by road and rail I. A. Wood	183
Chapter 10	Domestic and Industrial Storage H. Jagger	197
Chapter 11	Discharges from industrial plants and the like into sewers, rivers and the seas G. F. Oldham	213
Chapter 12	The human factor in oil pollution Prof. W. T. Singleton	233
Chapter 13	The role of law in the prevention of oil pollution Prof. E. D. Brown	253
Chapter 14	Conclusion J. Wardley-Smith	295

Sources and acknowledgements   303

Index   305

List of figures

1. Relative size of oil tankers 1886-1968 — 3
2. a) Pollution of wells and ground water
   b) Electro-osmosis preventing oil from entering water extraction point.
   c) Electro-osmosis used to remove pollution — 21
3. Schematic diagrams of exploration and production wells — 66
4. Blow-out preventer stack — 69
5. Christmas tree — 71
6. Sub-surface safety valve — 72
7. World movement of oil at sea 1960-1975 — 80
8. Comparison of sizes of 250,000 DWT and 1,000,000 DWT tankers. — 80
9. Primary structural strengthening members and partitions in 250,000 DWT and 400,000 DWT tankers. — 81
10. Cost in dollars per barrel of transporting oil by tankers over a typical 11,000 mile voyage. — 83
11. Annual steam and motor vessel tonnage lost as percentage of total tonnage at risk — 85
12. Annual average loss ratios of total tonnage at risk by vessel type. — 86
13. Annual oil outflow (in tons of oil outflow per 1000 tons DWT) by tanker size — all categories 1969-1973. — 89
14. Stability and buoyancy in groundings — single bottom tankers. — 91
15. Stability and buoyancy in groundings — double bottom tankers. — 94
16. Positioning of ballast in tankers. — 98
17. Arrangement of inert gas distribution system — 100
18. Schematic comparison for emergency manoeuvres of large tankers. — 102
19. VTS scheme applicable to English Channel — 110
20. 32,000 DWT crude oil carrier using load-on-top system for anti-pollution — 122
21. Possible sources and remedies of oil pollution during port operations. — 134
22. Major pipelines in Western Europe — 154
23. Underground petrol storage tank — 205
24. Horizontal cylindrical storage tank. — 206

25.	Rectangular domestic storage tank.	207
26.	Longitudinal section of an API type separator	218
27.	a) Designs conforming to stereotypes	
	b) Designs encouraging misreading	235
28.	a) Ambiguous control/display relationship	
	b) Clear control display relationship	236
29.	Schematic representation of supervisory control.	249

List of plates

1. Oil on shore off Tierra del Fuego as a result of the *Metula* grounding. — 6
2. The wreck of the *Amoco Cadiz* — 10
3. Mussel rafts at La Coruña, Spain, after the *Urquiola* spill in 1976. — 12
4. Road tanker crash at Westoning, Beds. — 14
5. Stranded crude oil on the beach at Safaniya, Saudi Arabia — 25
6. Dead limpets after the *Torrey Canyon* wreck. — 27
7. Oil deposited by a high tide on Spartina grass of a salt marsh in Milford Haven. — 30
8. Automatic air sampling station. — 39
9. A power station, illustrating the tall chimney policy — 52
10. Battersea power station showing effect of washing of gases. — 56
11. Shiphandling simulation facility. — 101
12. View from the bridge on board the BP tanker *British Respect*. — 116
13. Lightening operation: *British Dragoon*, 53,000 grt, alongside a VLCC. — 127
14. Loading product at the Shell Iron refinery, Singapore, into the *Koratiya*. — 139
15. Remote controls on the tanker end of a chiksam (rigid arm, outflow boom) unloading a VLCC. — 140
16. Hydraulic flange coupling, which obviates the need to fit all the flange bolts. — 140
17. Aerial view of jetties at BP's Kent refinery. — 145
18. Semi-submersible lay-barge *Choctaw II*. — 163
19. Rail loading area at BP's Kent refinery. — 189
20. Loading a road tanker. — 189
21. Road tanker which delivers and meters domestic fuel oil — 192
22. Esso terminal at West London showing bunds. — 199
23. Crude oil storage tanks on Widdie Island, Bantry Bay, Eire. — 209
24. A biological percolating filter showing rotary distributor — 226
25. API separator, sand filters, and percolating filters in use at a refinery. — 231
26. The *Betelgeuse* after the explosion in Bantry Bay. — 296
27. Control room of the BP tanker *British Respect*. — 298

List of tables

1. World oil production — 2
2. Tonnage and number of world tankers — 5
3. Tanker accidents resulting in oil outflow 1969-1973 from tankers over 3000 DWT. — 10
4. Sources of oil in the marine environment. — 11
5. Serious casualties to tankers of 10,000 DWT and more by year and size of ship (1968-1976). — 87
6. Lives lost due to serious casualties to oil/chemical tankers 1968-1976. — 88
7. Types of equipment to assist in stopping of tankers — 107
8. Approximate reductions in stopping distance for various design changes for a VLCC operating at 6 knots. — 107
9. The sequence of anti-pollution operations in port — 146
10. Sample pollution prevention check list. — 148
11. Summary of the main equipment concerned with pollution to be found on board a tanker together with their design and testing requirements. — 150
12. Classifications of error. — 242
13. Accidents in offshore installations 1971-75. — 242
14. Job analysis methods and techniques. — 245
15. Limitations of human performance. — 247
16. Total error reduction strategy. — 250

1
Sources of pollution

J. Wardley-Smith OBE, BSc
*formerly Technical Manager
The International Tanker Owners Pollution Federation*

Man has always polluted his environment. Neolithic man, the hunter and food gatherer, living in a cave, threw the bones from his meal with broken pots and tools onto the midden at the cave entrance — pollution which, incidentally, we now use to identify his life style. This type of refuse was relatively harmless. There is, however, the story that, some three thousand years later, besieged medieval castles sometimes fell due to the fouling of their water supply by their own sewage and garbage.

The Industrial Revolution changed all this, as the amount of waste material brought to one place far exceeded what could be got rid of by nature. The manufacturing process was the be-all and end-all of the business. Any waste was piled on adjacent land or discharged into a river. The Yorkshire saying, 'where there's muck there's brass', meaning dirt and money go together, exemplifies this attitude of mind. In some cases the result was catastrophic; for example, in the Swansea valley, home of the metallurgical industry, fumes from the smelting of copper and zinc killed all vegetation around the works and the subsequent eroding of hillsides by heavy rain can still be seen a hundred years later. In 1880, one area in the valley, Llandore, was described as 'ugly with all the ugliness of grime and dust and mud and smoke and indescribable tastes and odours.' As this sort of thing was quite commonplace during the Industrial Revolution, it is of no surprise that early oil wells discharged oily water in their immediate neighbourhood. Oil and oil products, which increased rapidly in quantity, were just a source of additional waste material to be discharged at random.

Slowly over the years the health effects of industrial discharges and wastes, not only on the workers in the industry concerned but on the health of people living in the vicinity, has been realised and become a cause for public concern. The extent of this concern varies enormously from country to country and from industry to industry. In many places, the sea is still regarded as an infinite sink

Table 1: *World oil production*

Year	World production thousand tonnes
1910	44,262
1920	105,516
1930	206,306
1938	280,500
1950	538,470
1965	1,564,935
1974	2,882,570
1975	2,747,660
1976	2,962,350
1977	3,063,660*

* Preliminary figures.
Source: IP Information Service — Petroleum Statistics

or drain into which any amount of chemical or other waste can be discharged without concern. This opinion, too, is changing. Some sea areas, particularly closed seas like the Baltic,[1] are the subject of conventions signed by all riparian states which limits and controls the discharge of substances either directly or via rivers into that sea.

Mineral oil has been known for centuries. Natural seeps and bitumen lakes have long provided a valuable raw material. Oil was found in quantity in Pennsylvania, USA, in 1848 and twelve years later, 'Colonel' Drake drilled the first well. In 1850, a Glasgow chemist, James Young, discovered he could distill a substance suitable for lamps, which had previously used whale oil, from cannel coal and shale. The new mineral oil was even better and also provided a much needed lubricant for the machines of the Industrial Revolution, replacing tallow and lanolin. The first cargo of oil, in barrels, to arrive in England from America in 1861 was carried on the brig, *Elizabeth Watts,* while in 1886 the first tanker, the *Gluckauf,* was launched. This ship was of 2,307 tons. From this small start the world oil industry began. Though initially major producing areas were in the USA, by 1900 the East and West Indies, Burma, Russia and Romania had developed oil fields, followed later by Venezuela and Mexico, while in 1911 Persia started the group of Middle East oil fields. More recently, the

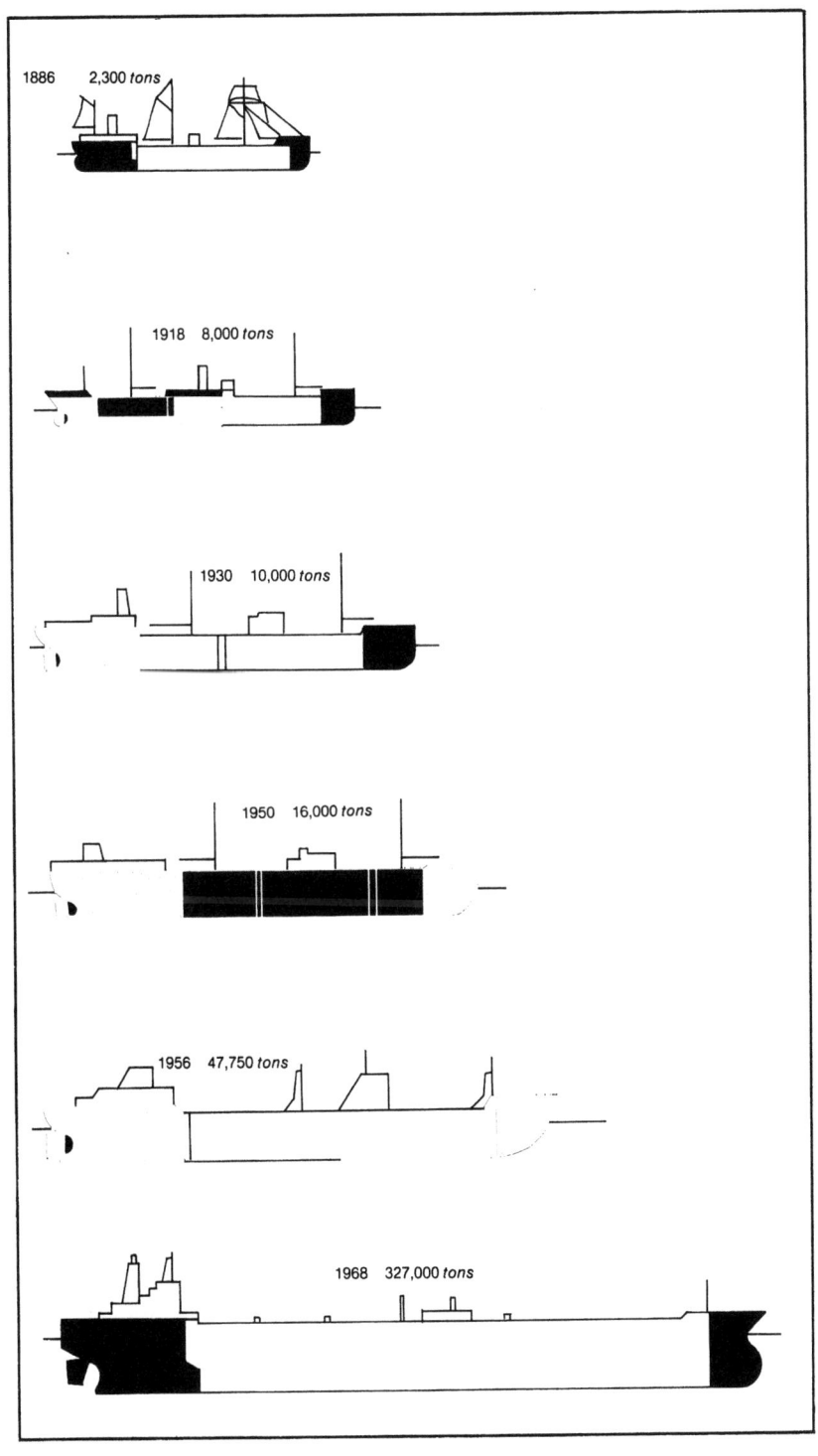

Figure 1: *Relative size of oil tankers 1886–1968.*

price and still growing demand has resulted in drilling in the seabed on continental shelves. As a result, the oil fields in the North Sea and on the West Coast of USA and many other sea areas are now starting production.

In general terms, excluding USA fields, oil has been found far from where it is wanted for use so that either the crude oil has had to be transported or, if refineries have been built close to the oil field, then the refined products have to be shipped or piped to where they are required. The growth of oil consumption has resulted in a similar growth in the oil tankers required to transport it (*see Figure 1*). There has been a steady growth in both size and number of vessels (*see Table 2*). The closure of the Suez Canal in 1967 made a very great difference. Previously, the size of a tanker was limited to the largest loaded tanker which could pass through the Canal. When tankers had to travel round the Cape of Good Hope from the Middle East to Europe the larger the tanker the lower the transport cost, and the limitation in size was given by the depth of water available in the loading and discharging ports, and by the depth of the sea available in certain areas of the route, for example the English Channel and for the Japan trade by the Malacca Strait. Tankers of 500,000 tons and over are sailing and on the drawing board. These immense ships have resulted in the construction of deep water transfer ports where the supertankers are completely or partially unloaded to shore. The lightened tanker can then proceed to its destination. These ports are found, for example, in Curacao, to supply the east coast of the USA with crude oil from Africa and the Middle East, and in Bantry Bay, Eire, to supply ports in Western Europe.

OIL POLLUTION

Oil pollution, particularly of the sea and navigable waters, has excited more public interest, nay, public concern, than any other waste or spilt material, even if the latter are potentially or actually far more hazardous. There are two simple reasons. Firstly, oil in general floats on the water surface and so can be seen even when in a very thin layer (the soap bubble colours of the thin oil film are produced by an oily layer only about one hundred thousandth of an inch thick, 0.3 microns). Secondly, any one walking on a sea beach or swimming in the sea is almost certain to have met the nuisance of oil lumps, often called tar balls, no matter in what part of the world he lives. In Western civilisation, there has been a huge

Year	up to 17,000 tons %[1]	17,000 – 25,000 tons %	25,000 – 50,000 tons %	50,000 – 100,000 tons %	200,000 – 250,000 tons %	Over 250,000 tons %	Total number
1951	80	13	7	–	–	–	2271
1961	29	23	43	4.7	0.3	–	4708
1966	11	18	36	31	4	–	5453
1971	3.3	10.2	19.9	27.2	33.8	5.6	6103[3]
1977	4.5		9.2	15.6	36.9	33.8	6912[4]

Based on deadweight tonnage[2]

Notes

1. From 1966 the percentages are for 10,000 dw and over, earlier for 2000 tons dw and above.
2. Dw tonnage is usually 50 per cent greater than gross, but with VLCCs it may be 100 per cent or more higher.
3. Total number for 1971.
4. Total gross tonnage 1977 is 174,124,444.

Source: *Institute of Petroleum World Statistics.*

Table 2: *Tonnage and number of world tankers.*

increase in the number of people owning and using small boats of all kinds for sport and recreation. Many, if not all, of these people will have experienced some degree of oil pollution of their paintwork at the waterline. It is not only the pollution of water by oil which causes concern, in many parts of the world every aspect of production, storage transport and use of both crude oil and of its derivatives are the subject of laws and regulations intended to reduce spillage and consequent hazards to human health and to the environment.

The properties of these materials can vary widely but there are a number of generalisations which can be made. Crude oil and petroleum products usually give off a flammable vapour which can be readily ignited. The primary hazard is therefore fire and explosion. This property gives rise to the regulations concerning all aspects of transport and storage and will be discussed later. These vapours are colourless and consequently invisible, and being heavier than air, they can flow from a spillage along the ground to accumulate at a low point or in drains from which they slowly diffuse away in still air. These vapours are themselves mildly toxic;

Plate 1: *Oil on shore off Tierra del Fuego as a result of the* Metula *grounding.*

low concentrations can cause anaesthesia whilst larger concentrations can cause suffocation by reducing the oxygen level. Liquid products, especially petrol and paraffin are poisonous if swallowed. Lead, an additive to most petrols, also increases the toxicity. Tank cleaning and repair is especially hazardous, so it is always carried out by specialised contractors. Most petroleum products are effective fat solvents and can cause dermatitis through prolonged contact with the skin.

Sea birds are particularly vulnerable to oil pollution. They encounter it on the water surface and some even seem to deliberately settle on oil slicks, perhaps because to the bird they resemble the indications of shoals of fish. The oil destroys the natural weatherproofing of the bird's feathers so that it loses its insulating properties and the bird dies of exposure and exhaustion. Oiled birds on the shore struggling to clean themselves are a piteous sight and so they too increase the disgust felt about oil on the sea.

The public considers that the oil floating on the sea must come from tankers. The occasional spectacular collision or grounding of these vessels encourages these views and, as tanker size increases, there is a growing feeling of alarm. In general terms this view is fallacious and so it is useful to consider, on a worldwide basis, where oil on the sea actually comes from, and if indeed spills are the only source of oil on the coastline and sea.

SOURCES OF OIL

In many oil fields, the strata overlying the oil have fissures which allow the oil to penetrate upwards and escape onto the surface. Oil fields can be both under the land and under the sea. Consequently there are a large number of natural underwater seeps and estimates have been made of the probable amount of oil which they lose into the sea each year. The figure probably lies between 0.2 and 6 million tons per year. Wilson in the USA gives as his best estimate 0.6 million metric tonnes per annum.[2]

In the early days of drilling for oil it was not infrequent for the oil pressure to be greater than expected and for the pipework equipment which was supposed to regulate the oil flow, to fracture and give rise to a 'gusher.' This accident is now very rare, but it still sometimes occurs. It has been estimated from USA experience that for every 1,000 wells drilled one may leak at some time, but in only one of every 10,000 wells drilled does the leak present any environmental threat. When drilling is taking place in water, particularly deep water, the difficult conditions increase the risk of a spill and make it much harder to stop. The Santa Barbara Platform B spill in California in 1970 was a classic example of an underwater spill. The Ekofisk spill in 1977 is perhaps the most recent major one. It is noteworthy that the total amount of oil discharged in both these spills was relatively small — under 100,000 tons in each case. In addition to the catastrophic oil well blow-out or spill there is always the possibility of small-scale leaks and the operational discharge of oily water and oily drilling fluids. In many countries these discharges are the subject of legislation and regulation. But oil is still accidentally lost. It has been estimated that the annual loss from exploration and production of undersea oil fields amounts to an average of 150,000 tons per annum.

The largest source of oil entering the ocean is from the land, either directly from effluent pipelines from refineries or petrochemical plants or from other discharges into rivers. These may be waste oil put accidentally or deliberately into the water course or the discharge of oily effluent from factories of all types. Automobiles use in total a great deal of oil. Some is burnt but some is discharged as oil mist and some drips from the vehicle onto the road or car park. A great deal of used sump oil is also poured into drains or onto the ground by car owners doing their own maintenance. Oil from any of these sources is likely to arrive eventually in the sea. Even gaseous discharges can be washed from the air by rain onto the ground and finally join the general run-off into the sea. The amount from this source is highly speculative. Some American authors have estimated several million tons. A more generally accepted figure for the run-off from land is 1.3 million tons per year. This figure can be reduced by the proper use of oil-water separators and the reduction of deliberate and accidental discharges. The design and use of separators is discussed later in chapter 11.

Discharges directly into the sea from tankers and other ships can be divided into four main groups:

1. operational discharges from tankers during tank washing;
2. bilge discharges;
3. spills caused by marine accidents, collisions, groundings, etc.;
4. spills during loading, discharging or bunkering.

Tank washing used to be the major cause of marine pollution from ships. If all the residual oil left in the tanks after normal discharge is washed out, then some 0.3 per cent of the cargo will be so discharged. Supposing 1.8 thousand million tons are carried by sea (author's estimates for 1974) then 5 million tons would be discharged.[3] This amount would be quite unacceptable so that, as discussed in chapter 6, several international conventions have been introduced and improved methods of tank washing (load on top and crude oil washing) have been introduced, which have greatly reduced the total amount of oil discharged in this way. The increase in the price of crude oil has given an added incentive to reduce the loss of oil from tank washing.

All ships take in small amounts of water which collect in the lower parts of the vessel or bilge. Oil fuel, used for firing boilers and oil used for lubricating can leak or be spilt and enter the engine room bilges which are periodically pumped out. Pumping of oily water into ports and harbours has long been prevented by local bye-laws. It is included in oil pollution conventions and legislation. Oil-water separators can and are fitted in most vessels. These are frequently not used when at sea. In total, oil from this source is estimated to be perhaps 300,000 tons per annum.

Ships of any kind are bound to have occasional accidents. People who live near a port, or who study shipping matters, are accustomed to see notices of vessels aground or colliding with other vessels. Seldom is anyone hurt and there is little 'news value' in the incident. It is true that when, more rarely, a vessel is perhaps lost in a storm the rescue of crew and passengers is headlined in all the media. If, however, an oil tanker, particularly a large one, has any form of accident, then, because of the possibility of oil pollution, it becomes international news. A survey has analysed the results of 450 accidents during 1969 to 1973 (it must be emphasised that many more accidents than 450 probably occurred);[4] it indicates that collison, grounding and structural failure are the major causes of the loss of oil *(see table 3)* accounting for more than 75 per cent of all the oil spilled. Nevertheless, considering the number of tankers and the number of very large ones (in 1976 395 tankers were of 240,000 tons deadweight or above), there have been relatively few large accidents. Some that have occurred have, however, been quite spectacular.

On 16 March 1978, the *Amoco Cadiz* carrying a cargo of 220,000 tons of light Arabian crude oil was rounding the Brittany coast on its way to the Channel and, while 8 to 10 miles off the coast, there was a rudder failure and a tug was called in. Unfortunately, the weather deteriorated rapidly and the arrival of the tug took a little time so that when a line was finally secured to the tanker it parted and, eventually, the tanker grounded on the

Table 3: *Tanker accidents resulting in oil outflow 1969-1973 Tankers over 3000 DWT*

Accident Type	Number	%	Oil Outflow (long tons)	% of total
Breakdown	11	2.4	29,940	3.1
Collision	126	27.8	185,088	19.4
Explosion	31	6.8	94,803	10.0
Fire	17	3.8	2,935	0.3
Grounding	123	27.3	230,806	24.3
Ramming	46	10.2	13,645	1.5
Structural Failure	94	20.8	339,181	35.6
Other	4	0.9	54,911	5.8
TOTAL	452	100	951,317	100

Brittany coast. The crew were taken off by helicopter and almost immediately the ship started to break up, battered by enormous seas and gale force winds. By 24 March the ship had split in two, spilling almost the complete cargo. As it was impossible to salvage any of the oil, the remaining tanks were ruptured by bombing. A very large amount of the oil landed on the coast of Brittany.

What was potentially an even more spectacular incident was the collision between two tankers, the *Ven Oil* and the *Ven Pet,* off Port Elizabeth, South Africa, on 16 December 1977. These two ships, owned by the same company, were each of 325,000 gross registered tons. One was loaded, the other in ballast. Fortunately,

Plate 2: *The wreck of the* Amoco Cadiz.

Table 4: *Sources of oil in the marine environment*

	Sources	Tons
a.	Marine operational loss	
	Tankers	1,000,000
	Bilge discharges	300,000
b.	Marine accidental discharges from all sources	350,000
c.	Offshore	
	Production	150,000
	Natural seepage	600,000
d.	Land based discharges Refineries, petro-chem plants and waste oils	1,300,000
	TOTAL	3,700,000

the total oil loss was only some 30,000 tons and, although a considerable area of the coast of South Africa was polluted, it clearly could have been a very great deal worse. Collisions are not common, but, considering the sophisticated navigation equipment carried by tankers particularly the very large crude carriers, they should be impossible. Two other much smaller ships, also sister ships, collided almost under the Golden Gate Bridge in San Francisco in January 1971. They were the *Oregon Standard* and the *Arizona Standard*. Some 4,200 tons of heavy fuel oil was lost in this incident.

The first major tanker accident in the world was the *Torrey Canyon* off Cornwall in 1967, when 119,000 tons of Kuwait crude oil was lost. In 1976, the *Urquiola* struck an underwater obstruction while entering the harbour of La Coruña in Spain (a pilot was on board and it was daylight). It caught fire and some 60,000 tons of its cargo of 106,000 tons of crude was burnt or lost. A rather similar accident occurred to *Jakob Maersk* while trying to berth at Oporto in Portugal, its entire cargo of 80,000 tons of crude oil was either burnt or lost at sea. Another, potentially, very large incident was the grounding of the *Metula* in the Straits of Magellan in Chile (a pilot was on board). In this case there was no fire and salvage was possible, although 50,000 tons of crude were lost out of the total cargo of 210,000. Table 4 gives an estimate of the amount of oil entering the marine environment in a typical year (say 1976) from all sources.[3]

EFFECTS OF MARINE OIL POLLUTION

Oil pollution of the shore, in addition to the reduction of amenity, also affects marine and shore life and vegetation. These points are discussed in detail in a later chapter. There is also the effect on humans to consider, namely the possible risk to man from the consumption of sea food contaminated by oil-derived carcinogens such as the polynuclear aromatic hydrocarbons (PNAH). Sea food can become tainted by oil pollution so making it inedible. It has even been suggested that skin cancers could result from sitting on an oil contaminated beach. As the sitting would have to extend for weeks, if not months, this can, however, be discounted.

The possibility of carcinogenesis has been considered by a joint group of experts under three heads:[5]

1. the carcinogens (PNAH) are bio-concentrated in the flesh of the fish so that even a low pollution level in the water can build up through the food chain until a significant amount is present;
2. if there is a threshold level for carcinogens, this may be exceeded by this bio-accumulations; or, alternatively, if there is no threshold there may be no safe level of pollution;
3. the PNAH may cause carcinoma in the fish and the consumption of such lesions could present a health hazard to man.

Evidence of bio-accumulation is fairly easy to determine. But it was also found that fish, including shellfish which, being static, are more likely to be contaminated rapidly lose most, if not all, of

Plate 3: *Mussel rafts at La Coruña, Spain after the* Urquiola *spill in 1976.*

the accumulated oil when they are kept in clean water for a time. It is, apparently, not yet proved whether there is, or is not, a safe level, but evidence is also lacking of carcinoma being produced by ingesting even high levels of PNAH-contaminated fish. There is some evidence of increased levels of stomach cancer in Iceland where large quantities of smoked fish are eaten — this does contain PNAH resulting from the smoking process. The major problem seems to be associated with bi-valves which, however, form but a small proportion of a normal diet. Many foods contain some amount of PNAH, and the total eaten is greater than would be expected from contaminated fish. In other words, a health risk is, if not quite absent, very low indeed, even when fish makes up a major portion of the diet.

A further factor which bears on the above is that the fish are likely to taste and smell oily. Very low levels can be detected and losses due to fish being unsaleable due to tainting are well known. It must be said that in many cases these losses are due to customer prejudice rather than actual tainting. For example, during the *Urquiola* incident in La Coruña, Spain, the sale of fish from that port in Madrid fell sharply though the fish in question had been caught in the North Atlantic and North Sea weeks before the pollution had occurred in the port. Nevertheless, this sort of thing represents a considerable loss to the fisherman who may have to cease operations until the oily conditions have abated and the shellfish have had time to clean themselves.

OIL POLLUTION ON LAND

This introduction, so far, has perhaps suggested that the major oil pollution problem is pollution of the sea. This is far from the case; pollution of the land by petroleum products can be equally objectionable and it is more difficult to carry out clean up or remedial measures. Early oil wells were drilled, as has been said above, without much regard to the environment. Wells drilled in the Middle East were largely drilled in uninhabited desert regions, so discharges of oil and waste material on to the ground presented no problem. Later drillings, in more hospitable zones, and now in the seabed are done much more carefully as will be described in a later chapter.

In many countries, major land spills come from storage tanks or from leaks or breaks in pipelines. In the case of Europe, there were, in 1976, 18,000 kilometres of pipeline but only fourteen

spills were reported which lost only some 6 parts per million parts of the oil transported down the pipeline. Spilt oil entering water courses can be collected by the use of dams or booms, and then pumped away. Oil which seeps through the ground into water-bearing strata below ground can travel, with the water, large distances and hence pollute sources of potable water. Such water can taste oily and objectionable with very low amounts of oil (of the order of one part in 10 million). So the small amounts, particularly of petrol or light fuel oil leaking from underground storage tanks can produce very serious problems. Fortunately, such instances are rare, but they do indicate the need to exercise great care at all times.

Oil in tankers, travelling either by road or rail, can also give rise to pollution and even to loss of life. A recent petrol tanker incident partially burned down a village when its driver lost control while descending a hill. In America there have been a number of accidents with rail wagons catching fire. Petroleum products, after accidents of various sorts, spilled without fire, tend to run down into waterways often aided by fire crews who wash roads clean to reduce the risks of combustion. Thus, incidents of this kind can easily lead to water pollution problems.

Oil spills can occur in all sorts of circumstances, despite equipment, safety devices, codes of best practice and the training of the personnel concerned. The US Coast Guard from their careful listing of spills into navigable waters have stated that over 85 per cent of all spills can be attributed to a human error.[6] Another paper examining spills from tankers reached a similar conclusion.[7] The latter study found that the smaller coastal tankers had many more loading/unloading operations per spill than did the very large crude carriers, which in general were newer and better equipped. The coastal tankers crews get more practice in the operation, while when the VLCCs crew have to carry out the tasks, it is at the end of a long and boring voyage and they are, perhaps, more interested in home leave than in avoiding spillage. All of these factors should be examined to determine if the training of the crew, the hours worked and probably the design, construction and layout of all plant and equipment is such as to minimise the chance of a wrong or careless operation leading to an oil spill. Legislation can only stop gross and operational discharges and do little to prevent spills. For example, it is difficult to imagine a situation more fraught with law and regulation than the road transport system of the UK but road accidents are commonplace and deaths from road accidents very large.

Plate 4: *Fire damage resulting from a road tanker crash at Westoning, Bedfordshire. No one was killed.*

To reduce the number of spills, every aspect of the handling of oil must be examined to determine if the whole operation has been designed to reduce the possibility of a fallible human operator machinery error. Cross checking, both mechanical and human, must be doubled and redoubled. This area, which is described at length in chapter 12 is probably the best if not indeed the only way to reduce these accidental spillages.

This book is not intended to be just a list of the circumstances and places where oil can be spilt. Rather it is a guide to those situations where oil spills have been known to occur. Those responsible for the organisation and planning of that part of the operation can then ensure that every step has been taken to reduce the likelihood of a spill. Even so, as there always will be some oil spills, it is useless to wait until they actually happen. Where spills are likely to occur trained clean up and remedial crews must be available and contingency plans must have been drawn up. Equipment of all kinds must either be specially stockpiled or arrangements made so that it can be borrowed and transported to the scene of the spill with the minimum of delay. It is only by considering all eventualities and planning for them in considerable depth that the ill-effects of a major oil spill can be reduced and mitigated. It is hoped that this book will indicate to people responsible for these actions the places and situations where spills are most likely to occur and consequently where they must take action *before* the spills occur.

REFERENCES.
1 The Helsinki Convention on the Protection of the Marine Environment of the Baltic Sea Area, March 1974.
2 J. Wilson, *Imports, Fates and Effects of Petroleum in the Marine Environment. Vol. 1* (Ocean Affairs Board, National Academy of Sciences, Washington D.C. 1973.)
3 J. Wardley-Smith, 'Oil Pollution of the Sea — the Worldwide Scene.' Regional Marine Oil Pollution Conference, Queensland, Australia, 1976.
4 J.C. Card, P.V. Ponce, W.D. Snider, 'Tankership Accidents and Resulting Oil Outflows 1969-1973.' Conference on Prevention and Control of Oil Pollution, San Francisco, USA 1975.
5 Joint Group of Experts on the Scientific Aspects of Marine Pollution — GESAMP — *Impact of Oil on the Marine Environment.* Reports and Studies No. 6 (FAO, Rome, 1977) p.91.
6 Rear Admiral W.M. Benhent USCG. 'Reasons for Collision and Groundings.' Safe Navigation Symposium, Oil Companies International Marine Forum, Washington DC, USA, 1978.
7 J. Wardley-Smith, 'Occurence, Cause and Avoidance of the Spilling of Oil by Tankers.' API/EPA Conference on Prevention and Control of Oil Spills, Washington DC, USA, 1978.

2

The effect of oil spills on land and water

A. Nelson-Smith BSc, PhD
Senior Lecturer in Zoology
University College of Swansea

Petroleum and its components which have been released into the environment are eventually degraded into simple compounds of their constituent elements by physico-chemical or biological agencies, with or without human assistance, and become innocuous; but, in the process, they may cause serious damage to plants and animals or their physical surroundings and thus impede human exploitation of natural resources. It is difficult to quantify or even summarise the effects of such pollution, since oils themselves vary so much. In laboratory tests of the toxicity of crude oils to winkles, Ottway recorded mortalities ranging from 1% to 89%.[1] Spillage of no. 2 fuel oil from a wreck on the Massachusetts coast had far-reaching and long-lasting consequences,[2,3] although a similar incident in southwest Wales involving oil of the same description but different specification had only minor local effects.[4] The effects of oil pollution also vary widely according to the history of the spillage, the nature of the locality and the state of its biota — Crapp found not only that the toxicity of a hydrocarbon-based emulsifier to two marine gastropods showed wide seasonal variation, but also that when it was most lethal to one it caused least damage to the other.[5] However, an attempt is made in this chapter to indicate the main consequences of oil pollution of the land, fresh waters and the seas, necessarily with some overlap or need to cross-refer. An oil pollution incident may interfere directly with industry or commerce, spoil the enjoyment of amenity pursuits or affect natural processes seemingly unconnected with human affairs. Different priorities may be assigned to remedial action accordingly, but it should be remembered that every influence which however remotely diminishes the richness and variety of our environment ultimately diminishes the fullness and perhaps even the span of our lives.

EFFECTS OF POLLUTION ON LAND

The nature of terrestrial oil pollution ranges from a massive single spillage — resulting, for example, from a split or overflowing storage tank, overturned transport vehicle or fractured pipeline — through the smaller, but perhaps repetitive, losses which often arise from careless handling at small factories and similar installations, or the surreptitious dumping of their waste oils, to such continuous but usually small-scale sources as an undetected leak or an oil-contaminated flow of waste water. Storm water carries from vehicle-parks and heavily-used roads a quantity of lubricating and other oils which is usually ignored, but which can be a significant source of local pollution. Although the site of any given incident cannot normally be predicted, most pollution is nevertheless restricted to the immediate vicinity of areas where petroleum is produced, processed, transported or used, since spilt oil rarely spreads far in the terrestrial environment.

Vegetation

The first noticeable effect of oil spilt across the surface of the land in any quantity is likely to be upon vegetation. Plants exchange the gases involved in respiration and photosynthesis through small pores, mostly on the underside of their leaves; some specialist plants of waterlogged, anaerobic soils also transport air from these pores to their roots, improving the soil condition locally.[6] The pores may readily be penetrated by thin oils, a process which is usually demonstrated by a darkening of the leaf as its air-spaces become filled with the oil; heavier fractions may block them up, while a coating of dark oil excludes or filters the sunlight necessary to the functioning of all green plants. Once it has received a significant covering of an active oil, an individual leaf invariably dies. Oil percolating into the soil around the roots may interfere with their uptake of water or cause the release of substances toxic to the plant. The damage which might be caused by a widespread spill of the more toxic or penetrating oils is indicated by the fact that selected blends have been used as herbicide sprays in their own right in addition to their frequent use as a medium for specific herbicidal substances.[7,8,9] Such spraying is usually intended to produce a total if temporary clearance of vegetation but it can be used selectively against weeds amongst a few crops, such as carrots, which are very tolerant of hydrocarbons. On the other hand, some light fractions have little toxic effect and have been used against plant fungal diseases or as solvents for insecticide sprays with little or no damage to crops or livestock.[10]

Fresh crude oil is, of course, a mixture of the potentially harmful light fractions and heavier, more inert material. Baker found that although shallow-rooted annuals on a salt-marsh were soon killed as a result of experimental spraying with a fresh crude, perennials with deep root-systems and sufficient food-reserves to enable them to produce new leaves were able to withstand eight or more monthly treatments.[6] In arid oil-producing regions, mobile sand around installations is often sprayed with crude or residual oils to stabilise it; vegetation soon breaks through, often at a greater density than elsewhere, probably because water-loss through evaporation is reduced. Heavy oil or bitumen has been used specifically as an agricultural mulch in parts of North Africa and the United States. In the process of 'hydroseeding', heavy oils are incorporated in the slurry of seeds and fertiliser projected onto areas otherwise difficult to cultivate, such as the scree-slopes on each side of a new road-cutting; the function of the oil is to bind the mixture, preventing over-rapid drying-out and also protecting the seedlings from being washed out by heavy rain before they have produced an adequate root-system. There have been numerous reports that petroleum contains substances which stimulate plant growth, but the effect is probably an indirect one on soil condition or favourable bacteria; Baker believed that those of her experiemental plants which grew better after oiling did so merely because flowering and seed-formation were suppressed, making more nutrients available for leaf and shoot production.

Vegetation in Arctic regions

Arctic or subarctic regions, which have recently become of great interest to the oil industry, present particular problems in the field of possible pollution damage. The vegetation is very specialised, most of the plants being sensitive to polluting crude oil;[11] dead foliage, bare soil or patches of standing oil resulting from a spillage absorb more solar engergy, melting the underlying frozen ground and creating an unstable muddy hollow which is not easily re-colonised. There is little variety, so that the elimination of even a single plant species has a much greater effect on the ecological balance than in a comparable temperate-zone habitat. Growth can be very slow; trees which may take 30-40 years to attain a height of less than 6 feet can be killed even without the spilt oil actually coating their foliage. Natural degradation of crude oil also proceeds very slowly under these conditions, but any attempt to remove it can cause further, long-lasting damage merely by mechanical disturbance of the surface.[12]

Animals

Damage to terrestrial animals will be confined mainly to those actually engulfed by the spill. Birds and small mammals may enter the polluted area to feed on insects or earthworms affected by the oil, becoming oiled or intoxicated themselves; wilted foliage may attract a few of the larger herbivores if it is only slightly tainted, although this is likely to be of only marginal importance. As a rather special case, heavy mortalities have been reported amongst ducks mistaking waste-oil lagoons and open oil-reservoirs for stretches of natural water while, of course, mortalities of the distant past in the La Brea tar-pits of California (a natural seepage) have provided valuable specimens for vertebrate palaeontologists.

FRESH WATER

Potable water supplies

The most serious effect is upon water supplies for both man and beast, especially by chronic pollution from an undetected leak or unauthorised waste discharge. Although oils do not penetrate completely waterlogged ground and migrate only very irregularly when the soil structure is not homogeneous, thin oils may move rapidly through some soils. The theory and characteristics of such migration have been discussed in detail by van Dam and by Dietz.[13,14] If ground-water supplies are drawn from nearby, drainage towards the point of extraction often entrains the oil. The three ways in which this can happen are shown in figure 2a. Well 1, situated on a perched aquifer, is at risk only over its small catchment area. The more typical well, well 2, into an unconfined aquifer is at risk from spillage over a wide area, while well 3, into a confined aquifer, is at risk from spills which can be distant from the well. It will also be seen that a river which has been polluted could also contribute pollution to well 2 if the aquifer is shrinking sufficiently rapidly. The standard method of dealing with such contamination is to sink relief wells upstream, although electro-osmosis can be used either to deflect or to concentrate the oil.[15,16] In figure 2b electro-osmosis is used to prevent the pollution from

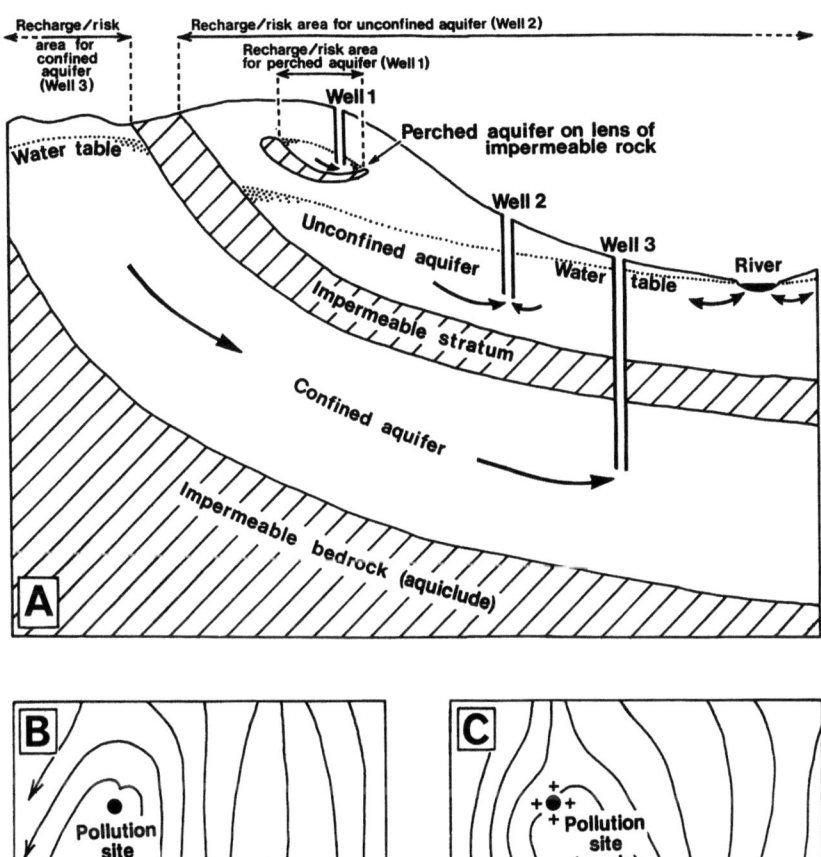

Figure 2: a. *Pollution of wells and ground water.*
　　　　b. *Electro-osmosis preventing oil from entering water extraction point.*
　　　　c. *Electro-osmosis used to remove pollution.*

entering the abstraction point and in figure 2c it is used to drive the pollutant to an abstraction point, which may be one of a number of wells specially drilled, and so remove the pollution. Human sensitivity to oily taints ranges from as little as 0.00005 ppm for gasoline with additives to between 0.2 and 1.0 ppm for crudes and 2.0 — 25 ppm for the more inert products such as lubricating oil, so it is unlikely that anyone could unwittingly drink harmful quantities,[17] but oil-contamination makes it necessary either to apply expensive special treatment or to waste a resource which is in increasingly short supply even in regions of good rainfall.

Rivers

Human settlements, industry and lines of transport are usually closely associated with waterways, so that many terrestrial oil spills almost immediately overflow into the fresh-water environment. Indeed, the first response to a road or yard spillage is often to hose it into the nearest drain or ditch. The larger waterways and lakes also carry oil-powered shipping and tank-barges. Unlike a spill on land, oil on water can be highly mobile, travelling far, under the influence of winds or currents, and spreading as it goes. A river has a constant, unidirectional flow which restricts damage to regions downstream of an oil-spill and soon flushes away the pollutant from any one reach; this makes it easy to locate appropriately the booms or other structures needed to divert or trap the oil. On the other hand, a heavy slick extending from bank to bank forms a barrier which most organisms may find it difficult to pass or avoid. Furthermore, if the upstream representatives of a particular species or community have been eliminated by heavy pollution, larvae or spores from lower down may find it impossible to move up against the flow in order to replace these losses. The inhabitants of a large lake or reservoir have more freedom of movement but, in the absence of such flushing, a massive spill or a succession of smaller ones can lead to exposure for a longer period and to the gradual accumulation of conservative substances. Whether the effect of a spill is mechanical or chemical depends on the relative proportions of oil and water, as well as on the nature of the oil. Gross pollution of a ditch, pond or small stream for any length of time is likely to be disastrous even if the oil is not especially toxic, as a result of the exclusion of light and oxygen, thermal blanketing and the overweighting or entanglement of small creatures and delicate structures. In a larger body of water, only the shallows and bank downwind, or regions where the current slackens, are likely to suffer such mechanical effects. Fresh, clean oils may well be repelled by the slimy outer covering of fish and submerged water-weeds, giving these some protection from minor spills, but waste oils often contain substances which enhance their dispersal or emulsification in water, while industrial effluents may include 'soluble' cutting-oils. These more readily cling to the surfaces of living organisms or enter life-cycles by becoming incorporated into sediments, shelters and food.

TOXICITY OF OIL

Straightforward oil pollution is easy to detect: in otherwise clean water, concentrations in an effluent which are greater than 10 ppm are visible as globules and even less oil will produce multi-coloured streaks on the surface. It is less easy to determine the presence of water-soluble components, yet any water which has been in prolonged contact with fresh crude oil or its lower fractions (as in a standard oil/water separator or interceptor) may have extracted significant quantities of potentially toxic hydrocarbons. Standard reference works list solubilities of 820 ppm for benzene, 470 ppm for toluene and 360 ppm for pentane in the temperature range 16-22°C.[18] Benzene and toluene are lethal to fresh-water fish at 10 to 400 ppm, according to species and conditions of test; naphthenic acids — minority constituents of crude petroleum which are even more readily leached by water — kill fish at 5 to 120 ppm. The relative proportions of oil and water, or the speed with which the water might be able to carry away such toxicants from a polluted locality, may therefore also be important when assessing possible chemical effects. Benzene and related aromatic hydrocarbons have certainly been used as herbicides in irrigation channels: at concentrations above 300 ppm, they are not only effective against water-weeds but also kill fresh-water crayfish. Elsewhere, aquatic vegetation may have positive value as an aesthetic feature, a feeding or nesting area for birds, a source of shelter and food for more fully aquatic animals, a protection against erosion of the bank or a commercial crop (for example, watercress or thatching reed).

Effect on fish

The toxic effect of various oil products in fresh water may be modified by other characteristics of this variable habitat: hardness, temperature and pH, as well as the presence of other pollutants.[19] A number of agricultural and industrial wastes, for example chlorhydrocarbons and phenols, are preferentially soluble in polar liquids and may be stripped from the water by oil globules, which are thereby made more harmful. Paradoxically, fish are attracted by the 'smell' of certain poisons, such as the phenols in gasworks discharges, and may thus be drawn into danger; in contrast, other minority components and by-products of petroleum are repulsive although not particularly toxic, driving fish away from their normal breeding and feeding grounds or blocking their ability to find food. In a salmon river, trout stream or popular coarse-fishing locality, this effect could bring about a serious economic or amenity loss. An incident excluding the fish from a considerable length of waterway might be even more serious for predators other than humans at certain times of year, for example during the breeding season of a nearby heron colony when relatively huge amounts of fish are consumed by the growing young. At an earlier stage, even small amounts of oil, carried back to the nest on legs or neck plumage, could inhibit the hatching of eggs or expose young chicks to chilling. Swimming birds are in danger as adults, too; few people can have failed to see pictures of oiled swans in the news media. Aquatic mammals — beavers, otters, muskrats etc — suffer in much the same way as waterfowl.

At a lower level on the ecological ladder, both young fish and the smaller organisms on which they and their parents feed are often more sensitive to pollutants than the adults but less able to escape them; where there is no fish food, it is evident that there will soon be no fish. Furthermore, it is the worms, snails, small crustaceans and insect larvae living on the bottom, rather than the more obvious midwater and surface dwellers, which scavenge the organic wastes and detritus which build up there. In the absence of such scavengers, many sluggish rivers would become anaerobic, acid and malodorous — useless not only as a natural habitat or fishing area but also for water-sport or even as a source of water for industry. The lower reaches of the Cuyahoga River in Ohio provided, until quite recently, an extreme example of such degradation, being officially listed as a fire risk because of the inflammable garbage and oil floating at the surface, and the methane produced from its bed. Rivers continue to flow, polluted or not, even if their topography may change. Lakes pass through a natural ageing process, in which they ultimately become filled in, and many forms of pollution speed this process; early stages of such acceleration have been recorded from the American Great Lakes into which the Cuyahoga and similarly abused rivers flow.

THE SEAS

Although the seas are large, they also support by far the most important mode of petroleum transport and the greatest single source of oil pollution, the ocean tankship. Most polluting incidents involving these vessels occur not in the open oceans but in narrow waters or in the approach to ports. Other sources, such as offshore production platforms and undersea pipelines, also offer a serious source of pollution in coastal waters. A single incident on the open sea might not, in itself, be too serious except in its contribution to the total load of pollution which the oceans have to accept; the explorer/anthropologist Thor Heyerdahl, marine scientists from the Woods Hole Oceanographic Institute and others have described how thoroughly the surface of the North Atlantic

Plate 5: *Stranded crude oil on the beach at Safaniya, Saudi Arabia.*

is covered with oil-blobs.[20] The oil, which may originally have formed a thin continuous slick, is subjected to various physico-chemical processes as it is gently agitated by wind and waves; it thickens, emulsifies and much of it will sink to the bottom where bacteria and larger organisms eventually complete the process of degradation. What reaches the shore is thus well-weathered, tarry and containing much extraneous material; it, too, will eventually disappear, although not before some of it has stained the clothes of holidaymakers or been trodden into the carpets of their hotels. Closer to land, even a small oil-slick represents a major threat to those birds which swim at the surface or dive beneath it. Tens of thousands of auks and sea-duck die each year as a result both of large spills and chronic oil pollution. Bourne has given a succinct account of the mechanism of death or injury:[21]

> If birds come into contact with liquid oil it soaks their plumage and destroys its waterproofing and insulating properties so that the birds are chilled and liable to die of exposure and exhaustion when their energy reserves are exhausted. If they inhale or swallow toxic components of the oil while attempting to clean themselves this may lead to internal damage to the respiratory, alimentary or excretory systems. When the oil has begun to lose its volatile and toxic components by evaporation and solution and begins to solidify it may no longer soak the plumage and cause internal damage if it is ingested, but the birds may still damage their plumage trying to clean themselves, while if they smear oil on their eggs it blocks the pores in the shell through which respiration takes place so the embryo is asphyxiated.

Seals and smaller marine mammals suffer much the same consequences if their fur becomes oiled; seals also seem particularly prone to blinding by oils.

Effects on fish and shellfish

In shallower waters, oil sinking to the bottom either naturally or after treatment may cover the feeding or breeding grounds of fish, smother colonies of bivalves and foul beds of kelp or sea-grasses. Nets or other gear passed through such patches will also be oiled. There is some disagreement about the actual extent of damage by oil pollution to sea fisheries, since many other factors are present; it is, in any case, impossible to guess at the magnitude of indirect or long-term effects. After a well-publicised tanker wreck, it scarcely matters in immediate commercial terms, as demand for fish from the area usually undergoes a marked if temporary decline. Fresh crude oil, or products consisting of its lower-boiling fractions, may release enough water-soluble toxins to harm organisms in the underlying surface layers; small droplets are also ingested by planktonic animals, including small crustaceans of great importance as fish-food, in whose natural fat-stores have been detected aromatic hydrocarbons of petroleum origin. Recent long-term studies of plankton in the North Sea and northwestern Atlantic Ocean have revealed disturbing downward trends for which increasing pollution is a possible explanation.[22] Adult fish and other active mid-water organisms can usually move away from a polluted area, although not necessarily to a site which is as suitable for them, but their young are more at risk, whether in the plankton or on the bottom. In the laboratory, cell-division (and thus multiplication) of the planktonic micro-algae which are vitally important to most marine life is inhibited by additions of crude oil to as small an extent as 0.01 ppm; their photosynthetic processes are affected from 0.02 ppm. At such levels, fish eggs begin to hatch irregularly or late, while the development of already-hatched young fish or larval crabs and lobsters becomes abnormal between 1 ppm and 100 ppm. Fin- and shell-fish acquire a marked and persistent taint from less than 1 ppm of many oils. Immediately off a coast exposed to vigorous wave-action, droplets of fresh oil are readily carried down to the filter-feeding organisms such as barnacles, other crustaceans, tubeworms and bivalves which occupy rocks in the rich sublittoral fringe, or the corals

Plate 6: *Dead limpets after the* Torrey Canyon *wreck.*

which form fringing reefs in more tropical waters, together with the crabs, echinoderms and reef-fish which feed on them.[23] Tank-tests in the laboratory suggest acute toxicities (for an exposure of 24 hrs or more) of 2 ppm — 2 per cent for planktonic plants and animals, 10 ppm — 1 per cent for midwater organisms and 22 ppm — 5 per cent for those living on the bottom. Crude oils themselves, the condition of the animals affected and the circumstances of the pollution differ so widely that it is meaningless to attempt greater precision, especially as these 'concentrations' represent the amount of oil added to the test tank — at the higher values, it cannot be truly dissolved. Comparative figures for refined products are 60-200 ppm for gasoline, 300-3000 ppm for kerosine and diesel oil, 1000 ppm — 10 per cent for heavier fuel oils and 3000 ppm — 20 per cent for lubricating oils.

When a freshly-spilt crude oil, or the water-in-oil emulsion formed in vigorous seas and now widely known as 'chocolate mousse,' is cast up in large amounts onto a beach, it will cover much of the area thickly enough to bring about the mechanical effects of overweighting, exclusion of light and oxygen or thermal blanketing already mentioned. Wind and surf may cast it to considerable heights against cliffs, while the tides tend to shift it towards the top of the shore. Many rocky-shore organisms show a surprising degree of initial tolerance when covered with undiluted crude oil for an hour or two — often no mortalities at all are recorded after they have been transferred to clean water — perhaps because of their ability to close up tightly against the more common hazards of the intertidal zone. However, the few long-term observations which have been made suggest that they may subsequently neither grow nor breed. Even an hour's build-up of emulsified or heavy oil on the fronds of seaweeds may add enough weight for incoming surf to tear them from the rocks; lighter oil may cling in a thinner, patchy coat but concentrations between 100 ppm and 0.1 per cent can inhibit the process of photosynthesis.

SHORE VEGETATION

Maritime plants at the head of the shore, whose value to humans ranges from the aesthetic (as with the cliff flowers and colourful lichens of the Cornish coast) to the strictly practical (for example, the grasses and other plants which bind sand-dunes, storm-strands or marshy slacks) are no more tolerant of regular or accumulated oil pollution than those of salt-marshes; the elimination of a key species could lead to marked topographical changes such as the blowing-out of a dune system. More tropical shores are often protected from surf by a fringing reef of coral offshore and a belt of mangroves along the edge of the land. Again, the elimination of either or both could have serious consequences.

On temperate shores of moderate exposure, a balance is maintained between the larger seaweeds and animals such as limpets and winkles which graze away their settling sporelings. Limpets seem to be particularly sensitive to oil pollution and it is now a familiar observation that, following a serious oil-spill, the affected shores become over-grown with annual green algae, often followed by the perennial brown seaweeds. This reduces the area of rock available for the settlement of barnacles and mussels, affecting in turn both their predators and their contribution of larvae to the offshore plankton. In Cornwall after *Torrey Canyon*, where a lengthy stretch of coast was affected not only by crude oil but also by the overgenerous application of highly toxic dispersants, reasonably complete recovery took at least seven years; pollution of a much smaller area in Milford Haven by the less harmful product gasoline resulted only in a 'green phase' and recovery was rapid.[24]

Estuaries

An estuary, although it seems topographically just the mouth of a river, is in biological terms a branch of the sea. Whereas the river flows constantly seawards and most sites on open coasts are subjected only to the rise and fall of the tide, water movements in estuaries are more complex.[25] The tides give rise to a piston-like movement of sea water into and out of the lower reaches; plants and animals on their shores are thus subjected to strong currents in two directions, bringing marked changes in salinity as well as the normal problems of tidal emersion. A given block of water may not escape to the open sea until it has oscillated many times, either accumulating or distributing widely its load of pollutants.

As fresh water floats on salt, a net circulation is set up in which river water passing out to sea on the surface is balance by a wedge of sea water moving upstream along the bottom. Estuaries are valuable as nurseries for fish taken commercially further offshore, as well as for their own potentially rich fin- and shell-fisheries; they provide nutrients for adjoining coastal waters and feeding-grounds for many species of bird, especially over-winter. The demanding nature of their environment results in a low diversity of permanent inhabitants, although some species may be very numerous; many of these are tough and flexible but may nevertheless succumb to the imposition of further stress, such as serious pollution. An ecosystem of this sort is fragile, easily becomes unbalanced and subsequently recovers only with great difficulty.

Large estuaries are often exploited as sites for oil terminals, refineries, power stations and other heavy oil-consuming industry; they are thus at risk from such internal sources as well as pollution which may be introduced either from landward or seaward. More often than not, other forms of pollution (including waste heat) accentuate rather than diminish the impact of oil spillage. Bearing in mind both the delicate ecological balance and the greater likelihood of further pollution there, the consequences of oil pollution may thus be more serious in an industrialised waterway than on the open coast. It has been possible to monitor the ecological health of Milford Haven during its rapid transition from a relatively unspoilt deep estuary to Britain's largest oil-port.[26, 24] Its continuing satisfactory state is due partly to an unusually rich flora and fauna, the relative lack of other pollution and good flushing by strong tides; but the main reason is an enlightened and efficient anti-spill organisation. The essence of such a scheme is to maintain detailed contingency plans whilst being prepared to deal flexibly with the particular problems raised by each individual incident. Any decision on the treatment appropriate to a given spillage (in which, of course, one option is to do nothing) has to be a compromise between the conflicting demands of industry, commerce, amenity and conservation; it should be noted that, in the biological sphere alone, the requirements of (for example) commercial fishing interests, ornithologists and coastal conservation authorities may differ widely.

Plate 7: *Oil deposited by a high tide on* Spartina *grass of a salt marsh in Milford Haven.*

EFFECTS OF CLEANING UP THE OIL

Methods of dealing with coastal oil pollution which are acceptable in some circumstances may have serious biological consequences if applied in others, even when they seem the most effective in a purely mechanical sense. As rather obvious examples derived from the matters discussed above, sinking an oil-slick in the mouth of an estuary may not only foul a productive bottom but might also result in the oily mixture being carried upstream by the 'salt wedge' rather than out into deeper water; dispersing it into surf over a shallow rocky bottom or fringing reef could render it much more readily available to the ecologically important filter-feeders there; while mechanical removal of stranded oil from sand-dunes or salt-marshes is far more damaging than leaving it strictly alone. Such considerations are not immediately apparent, especially when under the pressure of an actual incident: consultation in advance with as wide a range of specialists as possible is thus essential in formulating any anti-oil-pollution plan.

REFERENCES

1. S.M. Ottway, *The ecological effects of oil pollution on littoral communities*, E.B. Cowell (ed) (Institute of Petroleum, London, 1971) pp. 172-180.
2. G.R. Hampson & H.L. Sanders, 'Local oil spill.' *Oceanus*, 15 (1969), pp. 8-11.
3. M. Blumer, G. Souza & J. Sass, *Hydrocarbon pollution of edible shellfish by an oil spill.* (Woods Hole Oceanographic Institution, Tech. Rep. Mass.) 70-1.
4. S.M. Ottway, 'The *Thuntank 6* spill.' *Annu. Rep. Oil Pollut. res. Unit Orielton*, 1971, pp. 29-38.
5. G.B. Crapp, *The ecological effects of oil pollution on littoral communities*, E.B. Cowell (ed), (Institute of Petroleum, London, 1971) pp. 129-149.
6. J.M. Baker, *The ecological effects of oil pollution on littoral communities*, E.B. Cowell (ed), (Institute of Petroleum, London, 1971) pp. 21-32, 62-77.
7. W.H. Minshall & V.A. Helson, 'The herbicidal action of oils.' *Proc. amer. Soc. hort. Sci.*, 53, (1949), pp. 294-298.
8. H.B. Currier & S.A. Peoples, 'Phytotoxicity of hydrocarbons.' *Hilgardia*, 23, (1954), pp. 155-173.
9. J.M. Baker, 'The effects of oils on plants.' *Environ. Pollut.*, 1 (1970), pp. 27-44.
10. L. Calpouzos, 'Action of oil in the control of plant disease.' *Annu. Rev. Phytopathol.*, 4 (1966) pp. 369-390.
11. T.C. Hutchingson & V.A. Hellebust, *Oil spills and vegetation at Norman Wells, N.W.T.* Task Force on Northern Oil Development rep. no. 73-43, (Information Canada, Ottawa, 1974).
12. K. van Cleve, *Recovery and restoration of damaged ecosystems*, pp. 422-455. J. Cairns, K.L. Dickson and E.E. Herricks, (eds) (University Press of Virginia, Charlottesville, 1977) pp. 422-455.
13. J. van Dam, *The joint problems of the oil and water industries*, P. Hepple (ed), (Institute of Petroleum, London, 1967) pp. 55-88.
14. D.N. Dietz, *Water pollution by oil*, P. Hepple (ed), (Institute of Petroleum, London, 1971) pp. 127-139.
15. J.C. Bruch & C. Taylor, 'The management of groundwater resource systems.' *J. environ. Planning*, 1(2), (1972) pp. 36-43.
16. J.C. Bruch & R.W. Lewis, 'Electro-osmosis and the contamination of underground fluids.' *J. environ. Planning*, 1(4), (1973) pp. 50-56.
17. J. Ineson & R.F. Packham, *The joint problems of the oil and water industries*, P. Hepple (ed), (Institute of Petroleum, London, 1967) pp. 67-108.
18. A. Nelson-Smith, *Oil pollution and marine ecology*. (Paul Elek Scientific Books, London, 1972).
19. L. Klein, *River pollution, vol. 2: causes and effects*. (Butterworth, London, 1962).
20. A. Nelson-Smith, *Petroleum and the continental shelf of north-west Europe, vol. 2: Environmental protection*, H.A. Cole (ed), (Applied Science Publishers, Barking, 1975) pp. 105-111.
21. W.R.P. Bourne, *The control of oil pollution*, J. Wardley-Smith (ed), (Graham & Trotman, London, 1976) pp. 72-78.

22   R.S. Glover, G.A. Robinson & J.M. Colebrook, *Marine pollution and sea life*, M. Ruivo, (ed), (FAO/Fishing News Books, West Byfleet, 1973) pp. 439-445.
23   E.J. Ferguson Wood & R.E. Johannes, *Tropical marine pollution*. (Elsevier, Amsterdam, 1975).
24   A. Nelson-Smith, *Recovery and restoration of damaged ecosystems*, J. Cairns, K.L. Dickson & E.E. Herricks, (eds), (University Press of Virginia, Charlottesville, 1977) pp. 191-207.
25   A. Nelson-Smith, *The coastline*, R.S.K. Barnes (ed), (John Wiley, London, 1977) pp. 123-146.
26   J.M. Baker (ed), *Marine ecology and oil pollution*. (Applied Science Publishers, Barking, 1976).

3

Air pollution from the use of petroleum

S.R. Craxford, ISO, MA, PhD, CEng, FInstF
lately Head of Air Pollution Division
Warren Spring Laboratory

To many people oil pollution of the environment means oil tanker wrecks, like the *Torrey Canyon* or the *Christos Bitas*; lumps of tar on the sea shore and perhaps local pollution in their neighbourhood caused by a spill or leak from an oil storage installation on to land or into a stream. The previous chapter has discussed the effects which oil spills can have on the living creatures and plants of the sea, the shore and the land, and mention has also been made of the possible harmful effects of oil on man himself. What is often not considered or thought of as oil pollution, and which is the subject of this chapter, is the air pollution that can be produced by substances emitted when petroleum is refined and its products burnt in engines and furnaces. The chapter also considers practicable methods of abating whatever pollution may occur..

To be certain that there is no misapprehension as to what constitutes pollution it is worth repeating the best of the definitions available, namely, that air is said to be polluted when the concentrations of some of its minor constituents are sufficiently high as to cause damage to health or loss of amenity. It is to be noted that loss of amenity includes unpleasantness, as distinct from adverse effects on health, caused by smoke, dust and dirt, malodour, diminution of sunshine, and so on. As far as man is concerned there is rather a fine line to be drawn between effects on health and on amenity as loss of amenity so often results in a state of mind that falls far short of the ideal of vigorous enthusiasm for living, even if it does not quite merit the description of a mental disorder. In addition to all this, adverse economic effects, such as damage to crops and materials, must also be taken into account.

The substances that are emitted into the air from the use of fossil fuels — whether coal or petroleum — and that give rise to concern, are:
- carbon dioxide
- carbon monoxide
- sulphur oxides
- nitrogen oxides
- smoke and smuts.

In addition to these, coal alone produces large amounts of grit and dust, and petroleum products are almost entirely responsible for the following:
- malodorous organic sulphur compounds
- hydrocarbon vapours
- photochemical pollutants, e.g. Los Angeles smog
- lead compounds.

All these substances will now be considered, with special reference to their status as pollutants. It is convenient to divide them into two groups — those arising mainly from motor engines, and those from stationary sources. The rather general problem of carbon dioxide stands separate from the rest and so will be dealt with first.

CARBON DIOXIDE

As some 1.5×10^4 million tonnes of carbon dioxide are released annually into the atmosphere by the burning of coal and petroleum, fears are recurrently expressed that this enormous quantity might cause trouble. In the natural course of events about 11×10^4 million tonnes are removed each year from the atmosphere by photosynthesis leading to the growth of vegetation and almost as much is returned to it by decay of the organic matter produced. Insufficient is known about the various processes involved to enable a confident prediction to be made as to the extent to which the extra carbon dioxide from the burning of fuels can be taken up into this natural carbon cycle, and hence to what extent increased carbon dioxide concentrations in the atmosphere may be expected. However, measurements have shown that from 1958 to 1963 the concentration in the atmosphere has increased by 1.3 mg/m^3 annually, which corresponds to about one-third of the amount actually released by the burning of fuel.

Carbon dioxide absorbs radiant energy in the infra-red so that its presence in the atmosphere in increasing amounts should reduce the loss of heat from the earth on these wavelengths. The average temperature of the earth should therefore be gradually increasing, and if this continued, the polar icecaps would melt, sea-level would rise and the major cities of the world, nearly all of which are low-lying, would be drowned. The fact is that the global mean temperature has actually fallen from 1950 to 1973 and is probably still continuing to do so, all of which shows that carbon dioxide is not the only factor, possibly not even a major factor, controlling temperature. It is, perhaps, worth remembering that the very large fluctuations of temperature that occurred during the Quaternary period, with tropical interglacial periods sandwiched between the major glaciations, occurred without any interference with natural processes by man.

It should, finally, be mentioned that there has never been any serious suggestion that the carbon dioxide in the air, even in towns, affects adversely the health or comfort of man.

CARBON MONOXIDE

When fuels — coal or petroleum — are burnt in engines or furnaces, all the carbon should be converted to carbon dioxide if combustion is complete and the maximum yield of energy is to be obtained. In practice, for one reason or another, combustion is often incomplete and part of the carbon is only oxidized as far as carbon monoxide, which is emitted into the air. Typical figures for the amounts so emitted per tonne of fuel are:

coal burnt in furnaces	0.5 kg carbon monoxide/t
oil burnt in furnaces	0.4 kg carbon monoxide/t
derv used in diesel engines	40 kg carbon monoxide/t
petrol used in petrol engines	300 kg carbon monoxide/t

These figures should be compared with emissions of carbon dioxide of the order of 3000 kg/t.

Unlike carbon dioxide, no global problem is posed by emissions of carbon monoxide, but the possibility of poisonous concentrations of the gas building up in the streets of towns has caused disquiet, possibly emotionally, on account of its reputation as a favourite method of suicide with a motor car engine running in a closed garage. The phrase 'poisonous concentrations' should be noted. Strictly speaking, there are no poisonous substances, but only poisonous concentrations of substances. For example, oxygen, the life-giving constituent of air, is highly poisonous at excessive concentrations, and most of the notorious 'poisons' are perfectly harmless at low enough concentrations. The poisonous action of carbon monoxide is due to its affinity for the haemoglobin in the blood, the substance which combines with oxygen in the lungs and transports it round the body. When carbon monoxide is present in the air it combines with the haemoglobin preferentially, leaving less for the transport of oxygen. An Expert Committee of the World Health Organization has examined the effects of carbon monoxide on health, and, on the evidence available to them, concluded that there was an increased risk for patients with cardiovascular disease if more than 4 per cent of the haemoglobin were put out of action in this way, and that this would occur if the air the patients were breathing continuously contained more than 29 mg carbon monoxide per cubic metre. For shorter exposures to carbon monoxide higher concentrations can be tolerated before the blood level rises to the limit just defined. The whole question is complicated, however, by smoking, as a heavy smoker may already have more than this 4 per cent blood level without exposure to further carbon monoxide in the air, and a lighter smoker will require much less than the 29 mg/m^3 of carbon monoxide in the ambient air in order to bring his blood level up to the 4 per cent limit. Thus the setting of a legal limit is a political problem fraught with difficulties. For example, should the cardiovascular patient who, contrary to medical advice, continues smoking, be protected or not, and then, what proportion of the non-smoking patients must be protected since there will always be a few, *in extremis,* to whom any stress will prove ultimately intolerable. But again, cardiovascular patients are unlikely to spend much of their time in busy streets where concentrations of carbon monoxide are at their highest. And on leaving a busy street, one has only to go some 20 metres down a side street for the concentration to fall to half its value in the main street.

Recent measurements at the kerb side in busy city streets in the United Kingdom[1] have shown that the concentration of carbon monoxide averaged over the working day will generally be within the range 5 – 9 mg/m^3, but with an hourly mean peak of 10 – 25 mg/m^3 at the rush hours. In streets of exceptional congestion the hourly mean peaks could reach 50 mg/m^3. These concentrations do not appear to be crying out for remedial action. However, if a limit is to be set to the concentration of carbon monoxide in streets, the connexion between this limit and the amounts of

Plate 8: *Site of automatic monitoring station for air pollution in London.*

carbon monoxide that may be emitted by individual motor cars is a tortuous one, and is not yet well understood, depending as it does on the design of the streets — width, direction, building height — on traffic density and flow, and on all the usual meteorological variables. The best that can be done is to introduce legislation on an arbitrary basis and of gradually increasing toughness, to reduce emissions, and to monitor whatever reductions in concentrations are brought about. It is very important that the amounts and the speed of the reductions in emissions be such that the motor industry can meet them without adding intolerably to the cost of motor transport.

Limits now in force for new vehicles in the United Kingdom and in Europe[2] can be met by a range of minor engine, carburation and timing adjustments at a very modest cost; indeed, fuel economy has probably been improved. The latest American regulations[2], however, can only be met by thermal or catalytic afterburners at a sacrifice of fuel economy, and the catalytic devices also require lead-free petrol.

No restrictions have been placed on emissions of carbon monoxide from stationary sources although these emissions are about equal to those from motor traffic. They are, however, made from high chimneys and are quickly dispersed, unlike the emissions from traffic which are at street level and tend to become trapped between the buildings on either side, and so making a major contribution to the pollution of the air in streets.

HYDROCARBONS

Like carbon monoxide, unburnt and partially burnt hydrocarbons will be emitted from motor engines in so far as the combustion is incomplete, and, on average, some 35 kg/tonne fuel are produced by petrol engines and 5 kg/tonne by diesels. These emissions give rise to mean kerb-side concentrations in busy streets of 3-4 mg/m^3, averaged over the working day, with hourly mean peaks of 7 – 10 mg/m^3 in the rush hours (exceptionally up to 15 mg/m^3). There is no evidence that these concentrations are harmful to man but they contribute to the unpleasant smell of traffic. Legislation[2] to limit emissions of hydrocarbons has been brought into force, since, under the right conditions, higher hydrocarbons in the air are essential to the production of photochemical oxidants and photochemical smog. The simplest requirement is that vapours that leak past the pistons of a petrol engine into the crankcase must be returned to the engine and not allowed to escape into the air. All motor engines are now arranged to comply with this rule. Otherwise the problem of limiting emissions is a question of improving combustion, just as for carbon monoxide, and the remarks made under that heading, in the last paragraph but one, apply equally to hydrocarbons.

NITROGEN OXIDES

Of all the oxides of nitrogen, the only ones that need be considered in the present context are nitric oxide (NO) and nitrogen dioxide (NO_2).

Nitric oxide is formed at the very high temperatures that occur in flames and combustion processes by the reaction of oxygen with the nitrogen that is a constituent part of some compounds present in oil and coal and with the nitrogen of the air; it is nitric oxide which issues from the combustion chambers of furnaces and internal combustion engines. As the gases cool some of the nitric oxide is converted to nitrogen dioxide, but this reaction soon becomes insignificantly slow on account of dilution as the gases emerge from the flue or exhaust pipe and mix with the surrounding air. As much as 10 per cent of the nitric oxide formed initially may be oxidised in this way. The total amount of nitrogen oxides emitted from stationary sources is probably some three times as much as that from traffic, but kerb-side measurements show four times as much nitric oxide as at nearby sites away from traffic so that pollution of the air of streets by this substance appears to arise predmoninantly from traffic. Peak values in busy towns are of the order of 1 mg/m^3 at the kerb-side. There is no evidence that such concentrations are harmful to man.

Nitrogen dioxide in the air arises from natural sources such as volcanoes, lightning and bacterial action in the soil; by the oxidation of nitric oxide by ozone; and from emissions from engines and furnaces as described in the preceding paragraph. A state of equilibrium is usually set up between these processes and the decomposition of nitrogen dioxide by ultraviolet light which gives nitric oxide and atomic oxygen, which in turn, produces undesirable oxidants (see below). The concentration of nitrogen dioxide finally attained depends on the meteorological variables and on the concentrations of nitric oxide, ozone and hydrocarbons.

The global background concentration of nitrogen dioxide is about 4 µg/m^3. By comparison, inner London has a long-term mean concentration of some 75 µg/m^3 with typical peak hourly averages of 200 µg/m^3. During summer high-pollution episodes hourly mean concentrations in excess of 450 µg/m^3 have been recorded.

Although there are insufficient epidemiological data to allow reliable air quality guides to be developed for nitrogen dioxide a World Health Organisation Group[3] proposed, on the basis of controlled studies on animals and human subjects, that a one hour exposure limit of 190 — 320 µg/m^3 should not be exceeded more than once a month. The Group considered such a limit to be consistent with the protection of the health of the public. They also considered that there was insufficient evidence to allow a limit to be set in terms of long-term averages.

These effects of nitrogen dioxide on health only constitute one of the many reasons for restricting the emissions of nitrogen oxides from motor traffic. The major cause for concern is the role of these oxides in the production of photochemical oxidants in the air, and these problems will be discussed in the next section before returning to the possibilities of abatement of the pollution they cause.

OZONE AND OXIDANTS

While the decomposition of nitrogen dioxide by ultraviolet light leads to the production of ozone, calculations made from the known rates of the various reactions involved showed that they could not account for all the ozone sometimes found in the air. It was then discovered that if certain hydrocarbon vapours, including some that are contained in motor exhaust gas and in various industrial emissions, are also present they take part in the reactions that give rise to ozone and could fully account for the concentrations observed. The extent of these reactions, and the concentrations of ozone produced, are greater the higher the temperature, the more intense the sunlight, and the greater the emissions of nitrogen oxides and hydrocarbons. Since some of the reactions are slow they are favoured by poor dispersion which helps to maintain the concentrations of the reactants. It takes several hours of irradiation of a mass of polluted air for the formation to ozone to reach a maximum. Conditions for photochemical pollution frequently occur for example, in such places as the Los Angeles basin in the United States, and in Mexico City, it is however only comparatively rarely that they obtain in the United Kingdom. While one hourly concentrations of ozone in excess of 160 $\mu g/m^3$ are normally limited, in the south of England, to irregularly spaced episodes of only a few days duration, things were very different during the heat wave of the summer of 1976. From 22 June to 12 July one-hourly concentrations of ozone remained between 175 and 500 $\mu g/m^3$ in the afternoons for the whole of the period throughout the southeastern part of England, in town and country alike. As will be seen later, these are very high concentrations indeed. It should perhaps be noted that during part of this time there was a steady drift of polluted air from industrial districts on the continent; but, for the period as a whole, home produced pollution from London and other towns in the region must have contributed significantly.

Up to this point reference has been made to ozone as the pollutant produced by these photochemical reactions, but in addition they give a number of by-products which, by their nature, must be taken carefully into account.

The first of these is nitrogen dioxide. As has already been explained it is the decomposition of nitrogen dioxide by ultraviolet light that produces oxygen atoms and puts in train the whole set of reactions that lead to ozone and the other photochemical pollutants. At a certain stage in the process, usually a few hours after the peak concentrations of ozone are observed, there is often a peak of nitrogen dioxide, lasting for an hour or two, with concentrations three or four times their normal values. For example, one-hourly averages have reached 280 $\mu g/m^3$ in the south of England, so that during the heat wave of 1976 concentrations were in excess of those regarded as tolerable by the World Health Organisation[4].

The second set of by-products are the organic oxidants, which consist of aldehydes, including acrolein, and, most notably, peroxyacyl nitrates (for short PAN), the commonest of which is peroxyacetyl nitrate $CH_3 {<}_{0-0-NO_2}^{0}$ All these compounds cause intense irritation of the eyes, nose and throat and act therefore as lachrymators. These, together with ozone, are usually classed together as oxidants, ozone being by far the most plentiful. In London and southern England in the summer of 1976 there was no report of PAN, which could hardly have been missed, on account of its characteristic smell, and there was no eye irritation.

The remaining set of by-products associated with photochemical pollution are the aerosols which reduce visibility. A characteristic feature of photochemical pollution in Los Angeles or Mexico City is the whitish mist that reduces visibility very seriously. This was again entirely absent in London during the 1976 episode although instrumental measurements indicated reduced visibility to a modest degree, when it was reduced to between 1.3 and 1.8 km. These figures indicate that the effect is about an order of magnitude less that in the American towns referred to. In Los Angeles the mist has been found, in part at least, to be an organic aerosol. It is now known that sulphur dioxide becomes oxidised in the photochemical processes that produce the pollution and a sulphate aerosol is formed which may be partly responsible for the reduced visibility.

While ozone itself is not a lachrymator it can adversely affect the respiratory functions so that, from the point of view of health, both ozone and the organic oxidants are a cause for concern. The Expert Committee of the World Health Organisation concluded that a one hour exposure to ozone, or other oxidants, expressed as ozone, should not exceed 200 $\mu g/m^3$ if the health of vulnerable groups of the population is to be protected.[4]

It should be made clear, however, that these limits assume the photochemical pollution to be of the type usually encountered in Los Angeles, with PAN present, whereas in the United Kingdom, when ozone concentrations are high, PAN is absent. At such times it is impossible to be sure whether distress among the old and infirm is brought about by the heat or whether it could have been caused by ozone.

In line with the conclusions of the Expert Committee of WHO, the United States National Ambient Air Quality Standard has been fixed at 160 $\mu g/m^3$ (ozone and oxidants, other than nitrogen dioxide, expressed as ozone) as a maximum one hour concentration not to be exceeded more than once a year.

In order to approach such a standard the main attack has been on emissions from motor engines. In principle, this is not easy as, according to one of the most fundamental laws of nature — the second law of thermodynamics — the higher the temperature of combustion in an engine the greater the efficiency, and so the greater the fuel economy. Unfortunately, while these conditions favour the reduction of emissions of hydrocarbons, they favour the production of nitric oxide. Control of emissions which lead to photochemical pollution has, up to the present in the United Kingdom and up to the last year or two in the USA , been based on a series of compromises in engine design and operation to lower the temperature of combustion sufficiently to reduce the formation of nitric oxide without too great an adverse effect on efficiency or on emissions of hydrocarbons. The latest regulations in the USA, however, cannot be met in this way and catalytic devices have to be used to remove nitric oxide, carbon monoxide and hydrocarbons from the exhaust gases. This adds inexorably to the price of the vehicle and, in addition, more expensive lead-free petrol must also be used as lead rapidly poisons the catalysts. While the annoyance and discomfort of Los Angeles smog provides complete justification for such steps to be taken in California, the question must be faced as to whether there are appreciable benefits to be gained by such a tougher stage of legislation in the United Kingdom or in Europe.

LEAD

As was noted above the fundamental law governing the operation of a heat engine, and which follows directly from the second law of thermodynamics, is that the higher the temperature at which heat is taken up by the working fluid, the greater the efficiency. In the internal combustion engine this temperature is increased by increasing the compression ratio. Using an ordinary type of petrol without additives a limit is soon reached on account of the tendency of the fuel to ignite ahead of the flame front advancing from the spark, and to give, in effect, an explosion which can be heard as a knock and which upsets the smooth running of the engine and ultimately damages it. This limited the compression ratio to about 4:1. It was found, however, that the addition of small quantities of lead alkyls (lead tetraethyl or lead tetramethyl, or mixtures of the two) suppressed this tendency to knock sufficiently to allow compression ratios of around 9:1 to be used, as is done at present, with a consequent increase of some fifty per cent in efficiency and a corresponding reduction in the amount of fuel used and in the volume of exhaust gases emitted.

Diesel fuel, which is a less volatile petroleum fraction and which is ignited in the engine by the high temperature produced by compression, needs no additions of lead.

The average lead content of petrol in the United Kingdom has varied downwards from 0.64 g/litre in 1966 to 0.45 in 1977. A legal limit of 0.45 g/litre has subsequently been introduced so that the average lead content is now probably arround 0.43 g/litre. Of this lead, about three-quarters is emitted with the exhaust as a mixture of complex halides which change in the atmosphere to lead oxide and carbonate dusts. All the particles that make up these dusts are in the respirable range and are fine enough not to settle out immediately. A World Health Organisation Group on Lead gave average concentrations of lead aerosol varying from 2 to 4 $\mu g/m^3$ in large cities with dense traffic down to 0.2 $\mu g/m^3$ in suburban areas and less in the country.

The remaining one-quarter of the lead in the petrol that is not emitted with the exhaust is trapped in the exhaust system or in the engine oil and filter. In addition, some 1 per cent of the lead alkyl in the petrol would appear to be lost by evaporation into the air at filling stations and from the carburettors of vehicles in motion, and concentrations of lead alkyl vapour are found in the air of towns amounting to one-tenth or less of those of lead aerosol.

A few years ago the U.K. Department of the Environment set up a Working Group on Heavy Metals, and this group reported on lead in 1974.[5] One of the questions dealt with was whether the concentrations of lead aerosol or of lead alkyl vapour found in city streets were harmful to man. Their answer was firmly in the negative. Special medical studies of policemen, bus drivers, petrol sales staff, and similar people failed to detect any ill effects that could be attributed to the occupational exposure of these subjects to lead from motor vehicles. This was in accord with the observed lack of correlation between concentrations of lead in the air and concentrations in the blood of men working there, and also with a whole series of estimates based on the intake of lead by the body from various sources, food and drink providing by far the largest items. These conclusions are completely in accord with those of the World Health Organisation Group on Lead, which reported in 1977. However, in spite of all this specific evidence, world opinion still tends to the view that it would be prudent to avoid an increase of lead in the environment, and that, where economically possible, it should be reduced.

Technically it is quite possible to reduce the amount of lead emitted by motor vehicles by altering the details of refinery operations so as to produce a petrol of a given quality without the addition of too much lead. For example, in 1965 5-star petrol (octane number 101) was made with 0.69 g lead/litre whereas by 1977 this was reduced to 0.47 g/litre, a reduction of 32 per cent while leaving the quality of the petrol unchanged. Another step that has been taken by the motor manufacturers in the United Kingdom has been to cease manufacturing engines that require 5-star petrol. This would seem, on the face of it, to be going rather far, unnecessarily, because, while in 1977 5-star petrol contained

0.47 g lead/litre, 4-star contained 0.46 g/litre, a negligible difference.

It is much more likely that the campaign against lead in petrol, while ostensibly being carried along on the emotional issue of the possible effects of lead emissions on health, is really more closely concerned with the question of avoiding poisoning the catalysts that are required to enable the most stringent standards for the purification of exhaust gases from carbon monoxide, hydrocarbons and oxides of nitrogen to be met. Such standards are already in force for new cars in the United States and pressures are bound to develop for their introduction elsewhere. In the USA it is a legal requirement that lead-free petrol must be available for use in these newer engines fitted with catalytic purification devices and it has been found that the provision of such petrol has not involved a prohibitive increase in price.

SMOKE AND SMUTS

Smoke is produced by incomplete combustion of fuel in furnaces and engines which leads to the emission of an aerosol of carbonaceous material. This may consist of anything from particles of almost pure carbon, through complex materials based chemically on condensed ring systems produced by cracking reactions at high temperatures, down to droplets of unburnt fuel. In principle, using oil fuels, there is no difficulty in avoiding smoke by proper design of combustion chambers and by careful operation, but in practice this is not always achieved. The worst offenders have been diesel engines as used in heavy road vehicles but boiler furnaces, both small and large, can produce this type of pollution.

The diesel engine works by drawing a fixed amount of air into the cylinder, compressing it to increase its temperature above the ignition point of the fuel, and then injecting the fuel as a spray. The power output is controlled by altering the amount of fuel injected. When the engine is running at moderate load there is little danger of its producing smoke, but as more fuel is injected and the fuel : air ratio approaches stoicheiometric any fault or imperfection in the design of the combustion system or in its state of maintenance that prevents perfect mixing of the charge, results in the formation of smoke, and finally, maximum power output can only be obtained by injecting so much fuel that heavy smoke is made. In the past there was always the temptation to increase the load of a vehicle, for economic reasons, to an extent that it could not be driven without making smoke, and even with the designed load, the driver was tempted to improve the performance on hills by using the excess fuel device that had to be provided for starting the engine, to inject enough fuel to increase the power but to make smoke.

In the United Kingdom[2] at the present (1978) there is a blanket legal requirement that 'every motor vehicle shall be so constructed that no avoidable smoke or visible vapour is emitted therefrom.' Specifically there are two other regulations controlling diesel smoke. The first of these states that when an engine is fitted with an 'excess fuel device' to facilitate starting, this device must be so arranged that it cannot readily be used while the vehicle is in motion if its use would bring about the emission of smoke. The second relevant regulation states that the engine must be of a type that conforms to the British Standard Specification BSAU 141a : 1971 as far as smoke emission is concerned (naturally enough this applies only to engines manufactured after 1972). The limits given in this Standard were based on what was considered acceptable in 1964 by a panel of ordinary people before whom a set of vehicles emitting different amounts of smoke were driven at constant speed. Since this legislation starts from the appearance of the smoke emitted by individual vehicles, it should be judged, in the first place, on similar grounds. In the opinion of the present writer, who travels a great deal by road throughout the United Kingdom, there is little or no appreciable remaining nuisance from diesel smoke, but there are those who do not agree and who would like a stricter limit.

However, quite apart from this nuisance aspect of diesel smoke, there remains the question of its contribution to the invisible burden of smoke in town air. Taking the United Kingdom as a whole, the total annual emission of smoke from stationary sources — domestic and industrial — is now about 0.5 million tonnes, and that from diesel vehicles about 0.1 million tonnes, so that, on average, the 16 per cent contribution from road traffic is not very important. But in certain special circumstances, for example, where there is a great concentration of traffic and where domestic smoke has been much reduced by clean air legislation, the proportion of smoke arising from road traffic will be well above the average. In London, for example, which is a prime instance of these conditions, recent estimates indicate that some three-quarters of the smoke in central districts, away from busy streets, has its origin in road traffic, and, naturally enough, a much higher proportion in such streets themselves. (These estimates are made in terms of the British Standard smoke scale,[6] as used by World Health Organisation in their recommendations).

The Expert Committee of the World Health Organisation, who examined the effects of air pollutants on health,[4] considered that in town air, which also contains sulphur dioxide, the health of bronchitics will not be adversely affected if the daily average smoke concentration does not exceed 250 $\mu g/m^3$. In 1976-7, the last year for which data are at present available, there were only two or three days in the year on which this value was exceeded at a few sites in central London; for most of the Greater London area there were no such days.

Smoke and carbonaceous dust is rarely a problem with modern oil-fired boilers and furnaces, and with such plant it can always be cured by attention to maintenance and operation. With older plant, just occasionally, mechanical separators — cyclones — may have to be used to prevent unacceptable emissions of carbonaceous dust. However, with any plant, a nuisance may be caused by acid smuts if the flue and chimney system is such that low enough temperatures occur on the walls for acid condensate to be formed. Carbonaceous dust is then trapped on the wet surfaces and pieces of the crust so formed become detached and are emitted as acid smuts which fall in the neighbourhood of the chimney to the intense annoyance of people living nearby. The remedy is correct flue and chimney design to avoid condensation, and also to take all possible steps to ensure rapid re-heating after the plant has been shut down.

SULPHUR OXIDES

Sulphur, in various forms, is a natural constituent of crude oil, and during the course of refinery operations it becomes distributed among the products, the latest figures for the United Kingdom being:

motor spirit	0.04 per cent sulphur
diesel fuel	0.35
gas oil	0.69
fuel oil	2.93

When these fractions are burnt, either in furnaces or in engines, all the sulphur is emitted as oxides. For the United Kingdom this amounts to 1.13 million tonnes (1975) and makes up about 43 per cent of the total sulphur emission.

Initially, some 90 per cent of this sulphur is emitted as sulphur dioxide and the rest as sulphuric acid aerosol, but as dispersion in the atmosphere proceeds the sulphur dioxide is slowly oxidised, either by photochemical or other processes, its half-life being about 5 days. Finally almost all is oxidised to sulphuric acid or sulphate (mainly ammonium sulphate) aerosol and is washed down by rain into the land and the oceans, to form part of the natural sulphur cycle. In this cycle sulphur from land and sea is converted to hydrogen sulphide by bacterial and similar actions and enters the atmosphere as such. There it is very rapidly oxidised to sulphur dioxide which, more slowly, is converted to sulphuric acid and sulphate aerosol which is washed down by rain as described above. It is considered that man-made sulphur dioxide contributes some 40 per cent to the total but the calculations are very uncertain. There are two stages of this sulphur cycle at which the man-made fraction might have an adverse effect on human health or economic prosperity. The first is in the immediate vicinity of the emitters where abnormally high concentrations of sulphur dioxide will be encountered before it is finally dispersed and diluted in the atmosphere at large, and the second is where the acid and sulphate aerosol is washed out of the atmosphere by rain.

Sulphur dioxide in towns and near industry

Up to this point, in considering the possible effects of pollutants on health, reliance has been placed on the WHO Report already referred to,[4] which is a helpful document for those pollutants that can be considered individually. It is, however, of necessity much less helpful for sulphur dioxide and smoke, as in town air these pollutants are always present together. The Report states that 'the worsening of patients with pulmonary disease' should be expected to occur at daily average concentrations of sulphur dioxide of $500 - 250$ $\mu g/m^3$; the corresponding concentration for smoke being 250 $\mu g/m^3$. It is stated that 'values for these two pollutants apply only in conjunction with each other' whatever this may mean. It is also specifically urged that these recommendations should not be considered independently of the accompanying text. Reference to this text indicates that the figures quoted are based on epidemiological work in London in the 1960s. Smoke and sulphur dioxide concentrations in this context are not independent variables and their ratio will vary from place to place and from time to time. The only valid con-

clusions that can be drawn from the London observations are the following set of 'either — or' possibilities:
1) smoke alone may be responsible for the observed effects on health, with sulphur dioxide playing no part, in which case the smoke limit holds without reference to the sulphur dioxide concentration, or
2) sulphur dioxide alone may be responsible, with the sulphur dioxide limit being valid without reference to smoke, or
3) some position between these two extremes may be the true one.

While there is no compelling evidence as to which of these various possibilities is correct, subsequent work has tended to point to some position very much nearer to 1) above than to 2). This weakness of the evidence against sulphur dioxide as a health hazard is emphasised in view of the increasing pressures to take very expensive steps to limit emissions further than are obtained by present-day practice.

At the moment, in the United Kingdom, there is a restriction of 0.5 per cent by weight on the sulphur content of gas oil, the fuel commonly used for central heating boilers and the smaller boiler plant in general. This serves to hold the basic sulphur dioxide concentration in the air of big cities to a reasonable value — annual mean concentrations below about 200 $\mu g/m^3$, usually below 100 $\mu g/m^3$. The remaining controls, of chimney design and height, are all left rather flexible so as to be adaptable to the special circumstances of individual installations. Although some of these controls are exercised by the local authority and others by the Alkali and Clean Air Inspectorate of the Department of Employment, they are all based on the principle made famous by the Alkali Inspectorate of 'best practicable means,' namely, that the best available means of abating pollution must be used that are compatible with technical and economic feasibility.

The first type of control concerns the height and situation of chimneys attached to buildings or placed near buildings in order to get the emission cleanly away from the site. As the wind blows over a building a volume of turbulence — the turbulent wake — is created in the lee of the building and gases from a chimney opening anywhere near roof level can be drawn down to the ground in this wake, where they can set up high concentrations of pollutants and cause a nuisance. Normally the chimney must be some two and a half times the height of the building to avoid this happening. However, in hilly terrain or near complexes of high

52. Prevention of Oil Pollution

buildings it is necessary to test the proposed scheme in a wind tunnel to see whether the height and situation of the chimney are likely to be satisfactory in this respect.

While controls on the lines described in the preceeding paragraph protect the immediate vicinity of a chimney, for example, the cottages around the factory gate, from undue pollution, rather more is required to protect a wider area of, say, 10 km radius around the site. It is a matter of everyday observation that sometimes the plume of gas from a chimney rises steeply into the upper air, sometimes it spreads horizontally at chimney top level for miles, and sometimes it rises a little and then drifts down to ground level at a greater or lesser distance downwind. There is also a complication of the gases sometimes forming a coherent plume but at other times breaking up into separate volumes moving in different directions. There is a multiplicity of formulae of varying

Plate 9: *A power station, illustrating the tall chimney policy.*

complexity which have been developed to describe these different types of behaviour in terms of different wind patterns and temperature gradients in the atmosphere, but all need the assignment of arbitrary values to one or more constants to make them fit the observations of concentration of pollutants downwind of the chimney. It is, in effect, impossible to calculate from first principles the average concentration of a pollutant at any given point in the neighbourhood of the chimney from which it is emitted, and the proportion of time this average is likely to be exceeded by a given amount. Two generalities do, however, emerge. The first is that the higher the chimney the less will be the concentration of pollutant at ground level and the further away from the chimney the point of maximum concentration, which occurs at a distance of some 15 times the height of the chimney, downwind. The second generality is that since the buoyancy of the plume depends on its temperature, that is, its tendency to rise like a hot air balloon, the hotter the emission the less the pollution at ground level. Based on long and detailed experience as to what conditions are apt to give rise to complaints and which are not, reviewed in the light of this type of theoretical and meteorological knowledge, official guidelines have been issued in the United Kingdom on chimney heights.[7] These recommendations do not apply to the very large plant such as the boilers of modern power stations, as meteorological conditions are different at heights of 250 m above ground from those that occur lower down, and modern large boiler plant is nowadays regularly provided with chimneys of this height, or more. A great deal of experimental work has been done in recent years by the U.K. electricity authorities on the behaviour of chimney plumes at such heights and the means of avoiding pollution from them is well understood.

Sulphur dioxide in the atmosphere as a whole

As explained above, the contribution of man-made sulphur dioxide to the global sulphur cycle is a further cause for concern. This arose in the early 1960s when the Scandinavian countries complained that the fish were dying in their lakes and their forest trees were being damaged on account of the acidity of the rain, as the sulphur compounds emitted into the atmosphere by the industrial countries of Europe, including the United Kingdom, drifted across their countries and were washed down by the rain. This complaint has since been investigated on an international scale, under the auspices of the OECD (Organization for Economic Cooperation and Development), but unfortunately the conclusions to be drawn from this work are by no means clear and no quantitative picture of the drift of pollution can be constructed. A proportion – but an unknown proportion – of the sulphur dioxide emitted annually in the United Kingdom certainly drifts over our coasts towards Europe, and of this, a small proportion will be washed down in Scandinavia. The investigation is now having to be replanned on a more satisfactory basis to obtain quantitative answers to these questions. On the effects side the position is equally uncertain. It would now appear that there is no firm evidence for damage to forests in Scandinavia by acid rain, and while there is an undoubted decrease in the fish population of the lakes of southern Norway it is still not proven that this has been caused by acid rain. These matters need settling as a matter of urgency as the very fact that they have been raised has led to international pressures to reduce emissions of sulphur compounds. If damage were proven, steps would have to be taken to reduce whatever emissions were responsible and all the possibilities are very expensive indeed. With this in mind a great deal of attention has been given in the last few years to the various options available.

The simplest solution for the United Kingdom would be to replace oil by natural gas but the provident conservation of natural gas reserves would probably rule this out, although estimates of reserves have a habit of increasing with time they still are a rapidly diminishing natural resource.

In the opinion of the present writer the most satisfactory of the remaining options is the desulphurisation, to the necessary extent, of fuel oil at the refinery. The technology is known and is of a type a refinery is geared to operate and that requires no expertise not already available there. The product would cost more than the fuel oil as at present used and would require more crude to produce a given amount. The next most satisfactory option also involves the removal of sulphur at the source, in this instance by gasification and desulphurisation of the gas, which is then used instead of oil as a fuel. Here again the technology is well known as it was used at gasworks throughout the United Kingdom before the advent of natural gas. It would provide a more expensive form of energy than fuel oil as now used and it would require the conversion of boilers and furnaces to burn gas instead of oil. It would have one advantage over all other possibilities in that complete desulphurisation of gas is easy and it could meet any requirement banning the emission of sulphur oxides from any particular installation. Both these possibilities have the added advantage that the sulphur product is elemental sulphur of high quality. This is now a readily saleable product but one that could easily be dumped without undue transportation charges if overproduction were brought about by increased demands for fuels containing less sulphur than at present.

The next set of possibilities are concerned with the desulphurisation of flue gas, to remove the sulphur from the plant after the fuel is burnt but before the products of combustion are emitted from the chimneys. The basic difficulty arises from the fact that the burning of 1 tonne of fuel oil produces some 15,000 cubic metres of flue gas, of which only about 20 consists of sulphur dioxide. Thus enormous volumes of gas have to be brought into contact with a reagent to remove the sulphur dioxide, a process that is slow and difficult on account of the low concentration of this substance. A large number of reagents have been proposed for this purpose and some of the processes based on them have been, and are still, used, mainly in America and Japan, on a full scale.

The simplest of these processes was that used at the London power stations of Battersea and Bankside, since 1931 and 1951 respectively, until just recently. The gases were washed with water from the Thames, to which a little extra lime was added, and the product - calcium sulphate - was discharged in solution into the river. Some 20 to 30 tonnes of water were pumped through the scrubbers for every tonne of fuel burnt in the boilers. Technically, the process was successful and removed from 70 to 90 per cent of the sulphur dioxide, but it did this at very considerable expense. Measurements of sulphur dioxide in the air of London provided no evidence, either when the processes were installed, or were out of commission, or were finally shut down, that they brought about any reduction in these concentrations. On the contrary, they gave rise to a recurring set of complaints of pollution arising from their use. Washing the flue gases with the enormous volumes of cold water, as mentioned above, reduced their temperature to about 50°C and they issued from the chimneys as cool wet plumes almost entirely lacking in buoyancy. These were apt to come down in the immediate neighbourhood, in Battersea Park for example, as wet malodorous clouds which, in fact, contained higher concentrations of sulphur dioxide than were present in their immediate surroundings. On days when the gases from part of Battersea power station were being washed and emitted from one chimney, and the gases from the rest of the station were being emitted from the other chimneys, the difference in behaviour of

the plumes was most striking, the one flattening out horizontally or falling quickly to ground level and the others rising high above the town.

This behaviour of a cool wet plume favoured the development of dry processes for the desulphurisation of flue gases but none of these were very successful. Nowadays technology in America and Japan tends to favour wet processes using a whole series of other reagents, followed by re-heating the whole mass of gas to restore its buoyancy. The cost of this final step can well be imagined. At the moment a number of wet processes are in operation in those countries to meet legal requirements, but the total cost is very high indeed. Although they can be made to work none are of a type that would appeal to a chemical engineer as a neat and elegant solution to the problem. In addition, the type of technology involved is foreign to a power station and the old jibe is not unjustified, that it would turn a power station into a chemical works producing electricity as a by-product.

There is another type of desulphurisation process that looks, at its present early stage of development, to be possibly more promising, namely fluidised combustion. A bed of sand or other such inert material is kept in a fluidised state by blowing air through it and when oil fuel is injected combustion occurs in the fluidised bed. By adding crushed dolomite or other alkali to the bed a proportion of the sulphur is retained. The spent alkali, together with a corresponding amount of the bed material, must be drawn off, as necessary, and disposed of.

If, ultimately, it is proved that sulphur compounds from the rest of Europe are responsible for the damage to the fish in the lakes of southern Norway and Sweden, a calculation will have to be made to see whether it would be cheaper to provide lime to put into the affected lakes to control their acidity rather than to instal and operate large numbers of desulphurisation plants in the rest of Europe.

MALODOUR

There can be no doubts about loss of amenity on account of malodour. Practically everybody objects strongly to the stink of motor vehicles in a traffic jam. Anyone who has lived within smelling-range of an oil refinery has objected to that aspect of its activities. This being noted, there is little that can be added constructively.

Plate 10: *Battersea power station. The plumes on the right show clearly the effect of washing.*

The smell of a diesel exhaust is at its worst under conditions that lead to the production of smoke, i.e. conditions of incomplete combustion, and is probably associated both with unburnt oil and with higher aldehydes. The former has a nauseating smell, and the latter are sharper and produce eye irritation. In the street it is naturally most noticable when a smoking diesel is passing, but this type of pollution may be worse with heavy slow-moving traffic when there is a temperature inversion and little wind so that natural ventilation is poor. The relative contributions of petrol and diesel engines to the smell of traffic is uncertain. It is probable that the modernisation of engine design will lead to very considerable abatement of this nuisance.

The mention of oil refineries at the beginning of this section was deliberately put in the past tense on account of the great improvements that have been made in the last decade. These have largely been brought about by attention to 'good housekeeping' on the plant, to deal with the leaking joint here and the faulty valve there, and so on. In the Rotterdam area this was helped by the recruitment of volunteers throughout the area, preferably people with experience in laboratories so as to be familiar with 'chemical' smells, who phoned the refinery whenever they smelt anything unusual.

CONCLUSIONS

The three most important conclusions to be drawn from the material presented in this chapter are the following.

1) There is no evidence that any of the putative air pollutants emitted in the United Kingdom from the use of petroleum is having adverse effects on health.

2) The reasons for the depletion of fish stocks in the lakes of southern Scandinavia should be carefully re-investigated, including the question of the drift of sulphur pollution from this country, as the evidence as it now stands is inconclusive. If the result should show that emissions of sulphur compounds ought to be limited the necessary technology is available.

3) Loss of amenity caused by the stink of motor traffic in traffic jams and similar circumstances should be borne in mind in the modernisation of engine design.

To these three one further fact should be added, namely that the replacement of coal by petroleum, as far as it has proceeded in the last two decades, has played its part in eliminating deaths of chronic bronchitics from exposure to coal smoke — and there were some 4000 such deaths in 1952 — and in cleaning up the dirt and grime associated with the use of coal.

REFERENCES
1   *Air Pollution from Road Vehicles.* National Society for Clean Air, Brighton.
    An authoritative statement of the present position being prepared for publication early in 1979.
2   *Clear Air Yearbook.* National Society for Clean Air, Brighton.
    An annual publication which contains an excellent summary of the current UK legislation as well as many other topics to do with the abatement of air pollution.
3   *Oxides of Nitrogen : Environmental Health Criteria No. 4.* World Health Organisation, Geneva 1977.
4   *Air Quality Criteria and Guides for Urban Air Pollutants.* Report of a WHO Expert Committee. WHO Technical Report Series No. 506. Geneva, 1972.
    This report is now being brought up to date and is being re-published in separate parts, by subject.
5   *Lead in the Environment and its Significance to Man.* Department of the

Environment. London, HMSO, 1974.
Discusses lead from motor traffic in relation to the total lead in air, water and food.
6 British Standard Specification BS 1747 : Pt.2. 1969
7 Chimney Heights. Clean Air Act Memorandum. Ministry of Housing and Local Government, 2nd Edition. London, HMSO, 1967.

FURTHER READING

The following titles provide a useful guide to further reading.

M.J. Suess and S.R. Craxford, (Eds.) *Manual on Urban Air Quality Management.* World Health Organization, Copenhagen, 1976.
   The introductory chapter is for administrators and is concerned with the construction of policy, the remainder is for their technical executives to help in putting policy into effect.

A. Parker, (Ed.) *Industrial Air Pollution Handbook.* McGraw-Hill, Maidenhead, 1978.
   An excellent reference book in its field; slanted towards British practice; rather heavy going in parts.

A.C. Stern, (Ed.), *Air Pollution.* Academic Press, New York, 3 vols.
   The standard text-book covering the whole field; slanted towards American practice; uncritical; heavy going in parts.

Chem Systems International Ltd. *Reducing Pollution from Selected Energy Transformation Sources.* Graham and Trotman, for the Commission of the European Communities, London, 1976.
   A very useful collection of data on emissions in the EEC countries, possibilities of reducing these emissions, with indications of costs.

4

Pollution occurring during exploration and production on land and at sea

D.R. Blaikley, BSc, PhD, MInstPet
Amoco (UK) Limited
Previously Environmental Co-ordinator, Amoco Europe

In considering oil pollution from exploration and production activities, usually abbreviated to E&P, one must, as in other oil-related activities, differentiate between oil pollution arising from operational and from emergency situations. There are also differences between operations onshore and offshore due to a large degree to differences in environment, the effects of oil on the ecology and the variations in equipment dictated by the environment. We must also differentiate between the various phases of the operation, exploration, production and certain aspects of transportation, although the prevention and risk of oil pollution from pipelines and tankers is considered elsewhere and will not be dwelt on in this chapter.

Before discussing the various possibilities for oil pollution and its prevention in E & P operations, it would seem worthwhile to look briefly at exploration and production activities since, in many cases, the preventive measures used are part and parcel of the operation itself. This is particularly the case where one is talking about the possibility of a blowout, its prevention and the means of causing its cessation.

DRILLING TECHNIQUES

The basic techniques of drilling exploration wells and producing an oil field once found are basically common to both onshore and offshore operations. The principle of drilling a well is fairly simple and is much as one would envisage. As in drilling any hole, a drill bit is rotated, in this case on the end of a long steel pipe, to cut a hole into the rock until sufficient depth is reached at which point it can be determined whether the rock formation, indicated as geologically favourable, does actually contain oil or gas. These depths can however be as much as 30,000 feet although more usually in the range of 10,000 — 15,000 feet. This means that the system has to be powerful enough to drill to this depth, whilst removing large quantities of rock cuttings, in addition the drill or bit must be lubricated. High pressures are encountered due to certain types of rock formation and if water, oil or gas are found. These pressures must be contained and the well or borehole made sufficiently strong to withstand their effects.

The most visible aspect of a drilling operation is the drilling rig. On land this has to provide the mechanical means of powering the cutting bit and handle the vast length of drill pipe by which the bit is rotated. In offshore operations the drilling rig has to not only provide the means for this but also has to store sufficient materials to keep the operation going for long periods at a time. In addition, an offshore drilling rig must provide accommodation for up to a hundred or so men, generate its own power and, in many cases, also be a seagoing vessel.

However, from the standpoint of understanding the risks of pollution it is the drilling of the well itself that is important and it is probably most logical to start with the drill bit. This is selected to match the size of the well and the type of rock being drilled. As the well is drilled progressively deeper the diameter of the hole is reduced in stages. The drill bit is made of steel and can have tungsten carbide or even diamond inserts to cope with the different hardness of the rock formation encountered. The drill bit is supported and rotated by the drill pipes to which it is screwed, the whole length being known as the drill string. At the bottom of the drill string are drill collars, 30 foot lengths of thick walled steel pipe which screw together and to the bottom of which the bit is attached. The collars provide weight and guidance to the bit where it is most needed. The drill pipe, also in 30 foot lengths screwed together, is attached to the drill collars and forms most of the drill string. The derrick is necessary to support the weight of the drill string in a controlled manner. It also provides sufficient height to withdraw the drill string from the well and enable it to be dismantled in 90 foot sections as it is withdrawn. The derrick must also be sufficiently large to allow these sections, each three lengths of drill pipe, to be stacked and then re-assembled for re-entry after, say, the changing of a cutting bit. Additional fittings, to give flexibility to the drill string or to stabilise it, can be fitted between lengths of drill pipe if necessary. At the rig floor a 40 foot length of square or hexagonal cross sectioned pipe, known as the kelly, is screwed into the last length of drill pipe. The kelly passes through a square shaped hole in the centre of a rotary table. This table is fixed in the floor of the rig and is turned to rotate the whole of the drill string. As the bit cuts the well deeper the kelly slides through the table until some 30 foot has been drilled when the kelly is unscrewed and an additional length of drill pipe is added.

To remove the rock cuttings and lubricate the drill string, drilling mud is pumped through a flexible high pressure hose attached to the top of the kelly by a swivel to permit rotation. The mud then passes down through the drill pipe and out through holes in the drill bit thus lubricating the bit. It returns up through the annulus between the wall of the well and the drill string. This flow washes out the cuttings and serves as a lubricant between the drill string and the wall. In addition the mud must be sufficiently heavy so that the weight of the mud column can prevent the oil, water or gas, under pressure in a rock formation penetrated by the drill bit, from flowing into the borehole. To achieve these functions the drilling mud is made up of barytes in a slurry with oil or water together with other additives such as Bentonite to seal the wall of the borehole. This prevents mud leaking out into porous rock formations, assists lubrication and helps to retain the rock cuttings in suspension. The density of the mud is varied by altering barytes concentrations to match the pressures expected. Mud chemistry is complex and the properties of the mud are varied to match different wells and as an individual well progresses. Since drilling mud is an expensive material it is continuously recirculated, being passed over fine mesh oscillating sieves to remove cuttings before re-use.

WELL PROGRESSION

In drilling an offshore well from a floating drilling rig the first step is to set a frame or template into the seabed, into this template is set a large diameter pipe from the rig floor. This serves as a guide for the drill string and is used to return the mud to the rig. Initial drilling to a depth of around 200 feet is with a large bit, say 30 inches in diameter, using water instead of mud. This avoids the weight of the mud column forcing its way out of the borehole through the rather weak rock formations at this depth. Before further drilling, a length of pipe, sometimes called surface casing, is cemented into the borehole to support the walls. Drilling then continues to about 1,000 feet where smaller diameter casing is cemented into the hole. Once this depth is reached low density mud is used as the drilling fluid and the well proper can be considered to have begun. Although the occurrence of rock formations containing high pressure fluids, oil, water or gas, is unusual at this depth, a blow out preventer may be fitted to the well at this stage. In offshore drilling this could be fitted at the seabed, onshore at the rig floor level. The BOP, described in detail when discussing the means of preventing a blow out, is a quick-acting device to close off the well should an incursion of fluid into the well threaten to overcome the weight of the mud column and displace it.

As drilling progresses the bit size is reduced and further lengths of casing are cemented into the well. The second length or string of casing is usually 9 5/8 inches in diameter, the final casing string being 7 inches.

WELL TESTING

The presence of oil or gas is usually indicated by pressure changes and from inspection of the cuttings. If such indications are found it will then be necessary to test the well to determine whether oil or gas are present in economic quantities. In addition, it is necessary to find out the characteristics of the reservoir to assist in its evaluation and further development as a potential producing field. Basically, this is done by permitting oil or gas to flow in a controlled and measured manner, from isolated segments of the well at the reservoir depth. The first stage is to ensure that the mud column's weight is adequate to counteract any formation pressures. With the BOP kept in place, the section of the formation to be tested is sealed off with a mechanical plug through which tubing is passed. Specifically sized holes are then punched through the casing, cement and into the formation using a series of electrically fired explosive charges passed through the tubing mounted on what is known as a perforating gun. This allows measurement of oil or gas flows from that zone. After testing of one zone, it will be sealed off, usually with cement, and a higher zone of the formation tested, this being repeated at a number of zones until sufficient information has been gathered. The well, after completion of testing, will either be filled with cement and abandoned or may be used for production.

WELL CONSTRUCTION

All exploration wells, be they on land or offshore, look very much the same. The only real difference is in the drilling rig, the manner in which the well is started and the interconnection between an offshore rig and the well head. The well casing in both cases serves to support the well walls, prevents incursion of fluids under pressure and also excursion of mud from the well into weaker rock formations. Offshore, the surface casing serves to prevent loss of mud into weak subsurface formations and thus into the sea, whereas onshore it often serves a second significant purpose in preventing contamination of ground water or underground aquifers.

Production Wells

In the drilling of a production well designed to produce oil or gas from a reservoir, the drilling operation is virtually identical to that for an exploration well. The main differences occur in the drilling of offshore production wells. In many cases the wells are drilled from a production platform fixed to the seabed, in which case the well head, together with the BOP, is on the platform and casing strings run from the platform into the seabed and on down. To maximise the extent of a reservoir that can be produced from an offshore platform, the production wells are usually deviated. This means that as the well is drilled downwards it is also curved away from the production platform to enter the reservoir several thousand horizontal feet distant from a straight vertical well. In this way a large number of wells can be drilled from a single platform to produce a large area of the field. The methods of drilling such deviated wells are sophisticated and the starting of such a deviation is generally achieved today by using a downhole motor. In such cases the drill bit is attached to a small turbine of similar diameter; this in turn is attached at a slight angle to the drill collars. The drill string is not rotated but the mud, on being pumped through the turbine, rotates the drill bit.

Historically, in the drilling of production wells on land it has not been necessary to drill deviated wells since the drilling rig can often be placed where desired and a vertical well drilled. However, today, there are often occasions when for environmental reasons, economics or in difficult geographical areas, it is desirable to drill as many production wells from one site as possible and here deviated production wells may also be used.

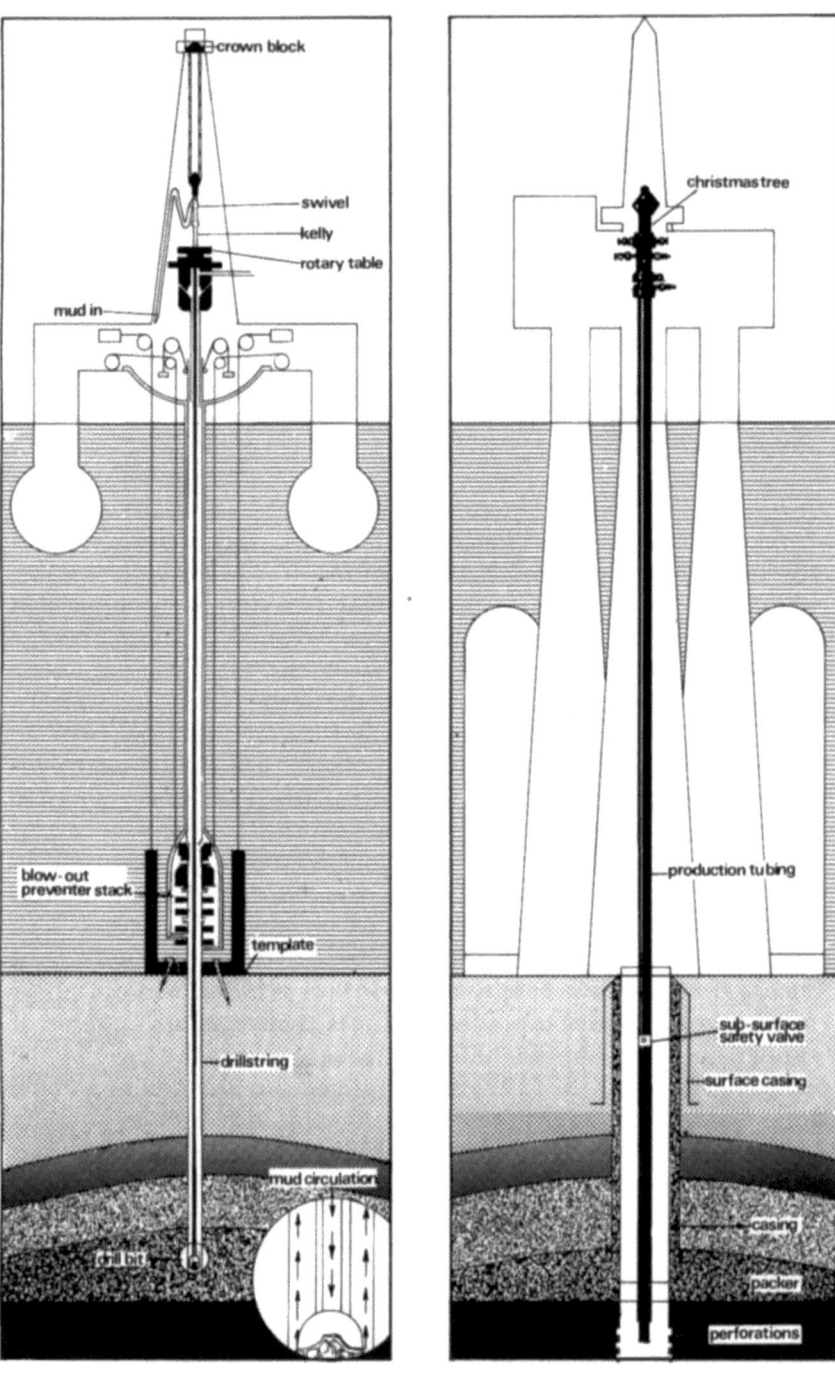

Figure 3: *Schematic diagrams of exploration and production wells.*

Once the well has been drilled and the casing strings set, then the production tubing is installed. This is narrow diameter steel tubing, again installed in screw jointed lengths of up to 4 inches diameter inside the casing and held in place by plugs or 'packers'. The space or annulus between this tubing and the casing is left filled with protective fluid and at the bottom is sealed off from the reservoir by cement and casing. With the well thus sealed off the BOP is replaced by a 'christmas tree'. (See p.70 for description of a christmas tree). A perforating gun is then passed through the christmas tree and the well perforated to allow flow of oil. The perforating gun, as other tools for serving a production well, is passed into the christmas tree via a lubricator to maintain the well sealed. In simple terms, the lubricator is a long tube with valves at top and bottom, in which the gun is held before being lowered into the well on a wire or rod which passes through a seal at the top. In this manner with the lubricator sealed at the top, the bottom valve can be safely opened to allow entry of the gun through the christmas tree into the well.

Production tubing is necessary to carry the flow of oil while yet protecting the casing from wear, excess pressure or other damage. The tubing string can be cleaned, replaced and used to support downhole safety equipment to shut off the well automatically if needed. The christmas tree valves can also be used to shut in the well either for an emergency or for maintenance, although their primary purpose is to control oil flow.

SOURCES OF POLLUTION

Having discussed briefly the methods of drilling and developing exploration or production wells, it is necessary to look into the matter whereby pollution can arise from an emergency and the methods used to prevent it. The worst possible occurrence, from the standpoints of both safety and the threat of oil pollution is, of course, a well blow out.

It must be stressed that contrary to many beliefs, such occurrences are rare, the worldwide average for offshore wells being two per thousand wells. Of these, some eighty per cent are gas wells and over sixty per cent of blow outs bridge or plug themselves by the collapse of the oil bearing rock under the flowing forces of oil or gas. This usually occurs within five to fifteen days.

In looking at the causes, means of prevention and means of controlling blow outs, it must always be borne in mind that a blow out not only poses a major pollution threat but also a threat to human life, tremendous material damage, loss of income, all of which combine to ensure that means of prevention and training are given first priority in any E&P operation.

A blow out is caused by the escape of oil or gas under its natural reservoir pressure instead of being retained during drilling by the weight of the mud column, by mechanical safety devices or during production by the well casing and the production system. Occurrence can be for a number of reasons. During drilling operations or during workover on production wells (where a well is opened up for maintenance work, etc.) their occurrence is primarily due to failure of final safety equipment following failure of the mud column to counteract the natural pressure of the reservoir after operational preventive measures. Preventive measures can be triggered by encountering an unexpectedly high formation pressure during drilling or, for example, by gas passing into the mud column and reducing its effective weight. Another possibility could be the loss of mud circulation where some drilling mud, instead of returning up the well, is lost into porous rock strata below the casing level.

None of these occurrences in itself mean that a blow out will occur since in virtually all cases the problems are countered by a progressive series of preventive measures. Such measures start with the initial geological survey of the rock formations to be drilled which are used to develop drilling strategy in terms of the rock bits to be used, casing lengths anticipated, mud types and weight best suited to the rock formations. During drilling many variables are monitored, the most significant from the standpoint of preventing a blow out can include the continuous monitoring of mud weight and volume, which is often computerised to ensure no reduction in mud density and no untoward decrease in the volume of mud in the well. Downhole pressures and the natural pressure of the formation being drilled are also measured. Often the latter is the best early warning system, assisted by the behaviour of the drill bit's rotation and the toolpushers' experience of any untoward problem, permitting early remedial action. This is through increasing mud density often combined with closing in the well and flushing out any oil or gas until an adequate mud column weight is restored. Problems of this nature as with further stages in well control, are dealt with as a matter of drilling practice by standard procedures developed on the job and in special training schools.

Well Safety Equipment

The measures discussed above are often known as primary control, for secondary control measures we must look at well safety equipment briefly mentioned earlier. Here there are significant differences between exploration and production operations, but in principle, practices are common to both onshore and off shore operations.

Safety equipment is primarily designed to seal off the well automatically if the weight of the mud column is not sufficient to balance the pressure of the reservoir thus losing control. Once the well is sealed, or shut in then control can be restored.

In exploration operations the main item of control is the blow out preventer. Several BOPs, each one designed to shut off the well under different circumstances are connected together to form a BOP stack (see figure 4). The BOP stack is connected to the casing string, the arrangement of the individual BOPs and their pressure ratings depend on the well, the state of drilling and the

Figure 4: *Blow out preventer.*

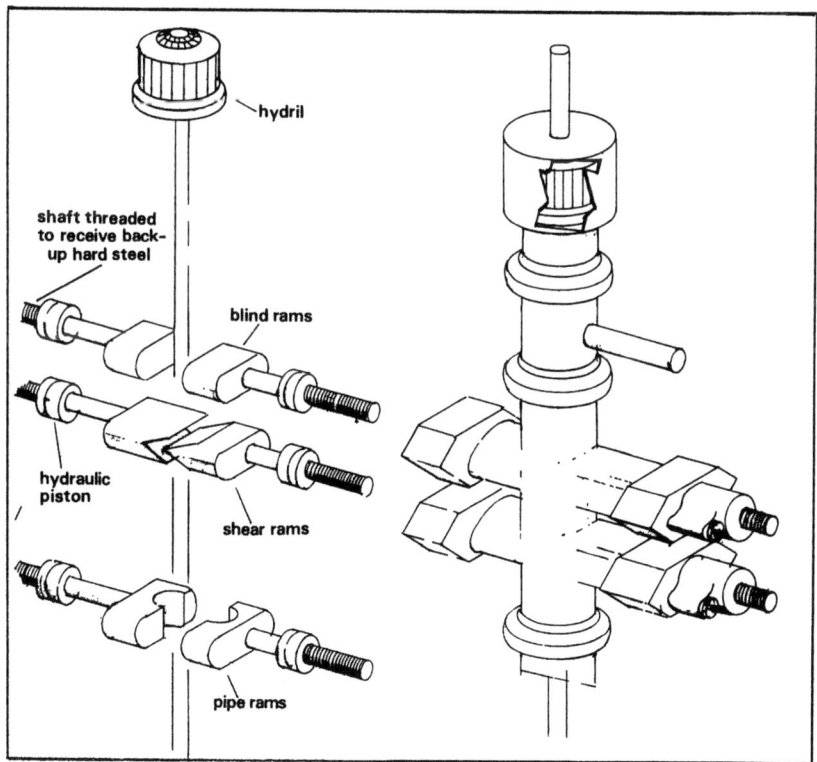

levels of problems anticipated. Two basic types of BOP exist; they are almost invariably used together in the formation of a BOP stack. They are the high pressure, ram type and the annular type for low pressure kicks and surges. In the ram type, which can be designed to handle pressures up to 15,000 psi or greater and can be readily adapted to serve a number of functions, each BOP has a pair of opposing rams which can be forced together. Pipe rams are designed to close round and hold the pipe when it is in the hole. Blind rams completely seal the hole if the pipe is withdrawn, while shear rams are designed to crush the drill-pipe sealing both the annulus and the pipe itself. Shear rams are placed above the pipe rams so that the latter can support the pipe after shearing. At the top of the stack is usually placed the annular BOP which can be used for sealing low pressure surges or kicks whilst permitting movement of the drill pipe in and out of the well and permitting rotation, both useful for, say, avoiding a stuck pipe.

In practice, BOPs are regularly tested and quite often used as an additional precuation should oil or gas pockets get into the mud column which then require circulating out. This is done by circulating the mud round to flush out the oil or gas whilst increasing mud density to counteract any formation pressures. Of course BOPs, although probably the most significant items of protection against a blow out during exploration, are not the only pieces of equipment used, and others, such as internal BOPs to fit in the drill pipe or valves to close off the drill string are also used.

In production operations the situation is somewhat different since the whole production system is designed to contain and control producing pressures and thus safety equipment is incorporated not only to control but to close in the well should problems at the wellhead or in the production system itself develop. The design of the wellhead and the production safety systems will, to a degree, be determined by not only the pressures involved but also the flows anticipated, the degree of well maintenance required and the type of oil, whether sweet or sour (i.e. hydrogen sulphide containing). At the wellhead itself the major item is the christmas tree, so called because of the numerous valves and side branches all of which have a part to play in ensuring safety and control of the system (see figure 5). Each branch or 'wing' is either a take-off point for the product oil or gas, or access into the casing annuli of which there are two or three. The latter provide access to change out casing fluid to assist in maintenance and to provide mud pumping points should well control be necessary if for example, the production tubing is withdrawn for maintenance.

Product is drawn off through the wing valve and then through a 'choke' assembly. The choke is a variable restriction used to control the oil flow to the treating system. All the casing valves, wing valves and chokes are duplicated to allow for wear, replacement and to provide a back-up. The master valve is the main valve used to shut in the well. It is only utilised after flow has been cut off by the wing valves to minimise wear of this valve.

Automatic surface safety valves are normally installed downstream of the master valve and are designed to shut in the well should temperature or pressure exceed certain specified levels. Particularly in offshore wells, but also onshore, two further types of sub-surface safety valves are used. The first, known as a 'storm choke', was originally used offshore in the Gulf of Mexico and designed to close in a well automatically should the platform or wellhead be damaged by a storm and result in uncontrolled flow. The storm choke is placed usually close to the bottom of the well and is activated by the flow of fluid or pressure, or a combination of the two, should certain preset levels be exceeded. Since these valves are difficult to test and can fail to close if the levels of flow

Figure 5: *Christmas tree.*

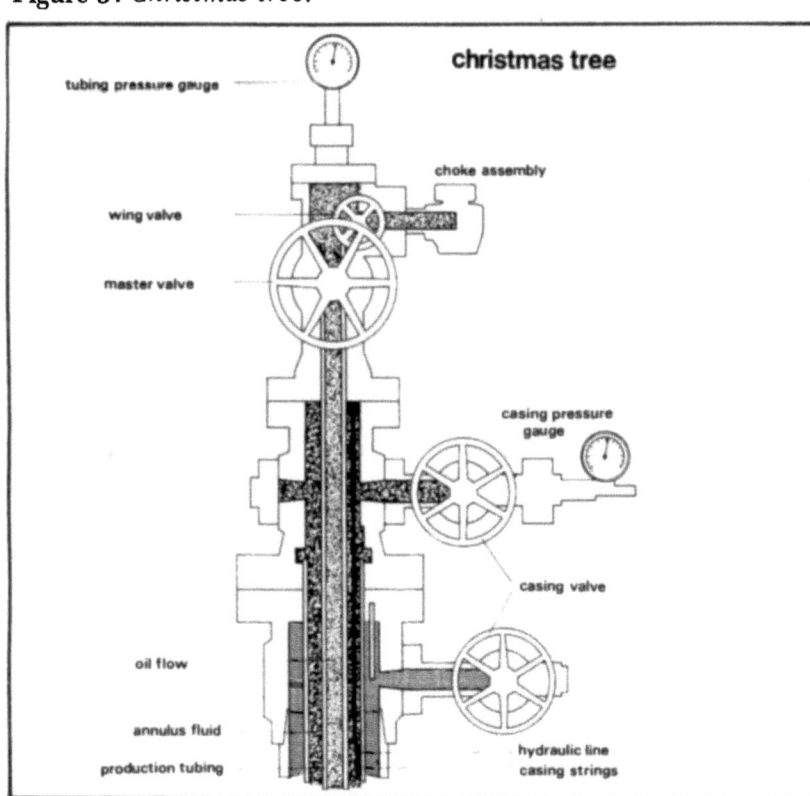

are not reached, for example partial blockage of the uncontrolled well by debris, they have largely been superseded by sub-surface safety valves.

Sub-surface safety valves are very simple in concept and are placed in the production tubing some 1,000 feet or so below the wellhead. They are serviced and tested regularly and have a very good record. The mode of operation of these valves is simple in that they are held open by hydraulic pressure from the wellhead area, or offshore, from the platform. If this hydraulic pressure is lost due to rupture of the line, manual operation, or in the event of a fire by the melting of a fusible link, the valve closes. A typical design is shown in figure 6 where in the open position the oil flows through a hole in the sphere. The sphere is held in this position by the hydraulic pressure compressing the springs. On loss of pressure the springs slide the ball assembly and in so doing rotate it shut. Another typical design uses a flapper valve, again held open against springs by hydraulic pressure. Loss of the pressure forces the valve to close and it is then held shut by the pressure of the oil from the reservoir.

Figure 6: *Sub-surface safety valves.*

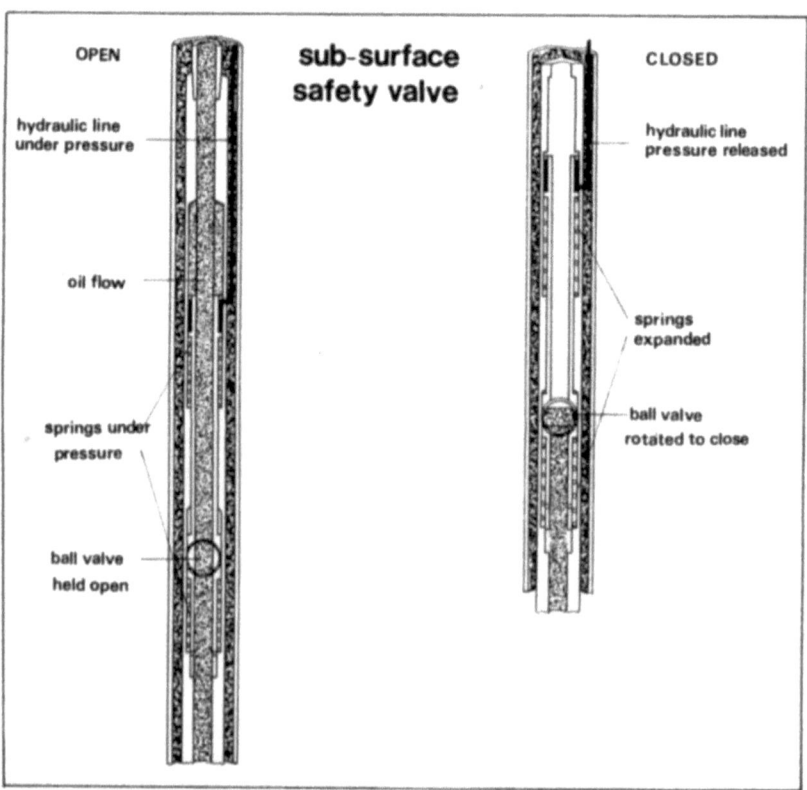

Apart from the equipment, there are of course procedural measures to minimise the possibility of a blow out. For example, on a production well there are occasions when the well has to be opened up for maintenance. This involves removing the christmas tree and installing a BOP stack. Here one goes through a number of stages to render the well safe prior to removing the tree. These can include firstly 'killing' the well by filling it with sufficiently heavy mud. Plugs are then inserted into the production tubing to 'latch' into pre-cut slots on the inside of the tubing to completely seal off the well.

One can obviously dwell in greater detail on the mechanical and procedural means of preventing a blow out and thus the most significant cause of oil pollution from E&P operations, however this is outside the scope of this chapter and is covered in numerous books specific to the subject. Despite the sophistication of the equipment, blow outs do occur for one reason or another, albeit infrequently, and as much from the environmental as other reasons it is worthwhile looking at the ways for bringing a blow out under control which is the primary method of limiting the extent of pollution.

RESTORATION OF WELL CONTROL

As stated earlier, if the reservoir rock is weak, or incompetent, i.e. liable to break up easily, say soft sandstone, the well will probably bridge and plug itself. Here the rock breaks up under the forces of the flowing fluid and progressively seals off the bottom of the well. Although statistically quite probable with weak formations, this cannot be guaranteed and consequently control procedures are initiated at once. To implement such procedures as rapidly and effectively as possible a contingency plan is brought into play. This has several primary objectives. The first is to mobilise those people required to implement all the necessary actions, ranging from pollution control, killing the well, to ensuring that the emergency work force has the necessary transport, supplies and food. Basic job functions and responsibilities are specified to minimise duplication and to ensure a co-ordinated reaction. Necessary emergency supplies, specialised equipment, its whereabouts and the means to obtain it are listed. The initial activities are outlined insofar as is possible bearing in mind each emergency will be an individual case and cannot therefore be covered in detail. It can thus be seen that a blow out contingency plan is similar to one for pollution control, but is far more extensive covering not only pollution but also the safety of the personnel involved and well control, they all being unavoidably inter-related.

Pollution control methods in the event of a blow out are identical to those for say a tanker spill. There is however one point which assists the operation, that is not having to deal with a massive once-off spill where all the oil is discharged at once, by virtue of a blow out being a continuing discharge, there is more flexibility in the reaction versus time to the incident.

In so far as regards the control of the blow out, or killing the well, there are basically two approaches, surface or sub-surface control. In surface control, the well is brought under control by capping. This is achieved by first obtaining access to the well followed by installation of either a large valve, a BOP or valve system similar to a single christmas tree. The first objective in such an operation is to obtain access to the wellhead. This could involve the clearing of debris, the removal of the existing BOP or christmas tree, if it is still in place or is damaged. In addition, it may be necessary, if the well is burning, to put out the fire and to keep the structure cool with water sprays to limit further fire damage; if the well is not burning it is necessary to limit the possibilities of ignition. Such operations, of course, are more difficult offshore where fire-fighting vessels and other drilling rigs or vessels may be necessary to provide access and working platforms. Once access has been obtained, the valve is installed by first securing it with a single bolt with the body of the valve out of the oil flow. The next stage, with the valve fully open, is to swing it across the flow and fully bolt it down prior to closing it to stop the oil flow. The same principle is applied whether a BOP or more complex system is used. Once flow has been stopped, drilling mud will be pumped into the well to bring it back under control.

In some circumstances, for example if the wellhead is too badly damaged, it may not be possible to cap the well and subsurface control must be undertaken. In this case a relief well must be drilled from a drilling rig at a safe distance from the blowing well. The relief well is drilled to break into the reservoir close to the bottom of the blowing well, with current techniques positioning of the well can be achieved to within 40 feet. It is only necessary to come close to the blowing well since further stages of the operation will open up flow paths between the wells. Firstly water is pumped down the relief well under very high pressure to force its way through the reservoir rock to develop these flow paths when the water will be carried into the blowing well. The water will then displace the oil or gas in the blowing well if sufficient quantities are pumped down, thus achieving 'flood out'. Once flood out has been obtained the drilling mud is pumped down the relief well to follow the flow path, filling both wells with sufficient weight of mud to bring the well under control. Once this control is obtained replacement valve systems are installed.

OIL SPILL REMOVALS

As has been mentioned earlier, the equipment and the methods of controlling a blowing well serve a threefold, inseparable function — to protect the integrity of the systems, to maintain, insofar as is possible, a safe system and to prevent the pollution that such an incident can cause. The procedures to remove the resultant oil are the same as would be used for an offshore tanker incident or an oil spill onshore. Similarly, precautions would be taken to contain or limit the damage caused, particularly when vulnerable ecological systems could be threatened. Onshore such precautions could involve bulldozing areas to contain the oil and thence pump it out, at sea removal of the oil from the surface by recovery equipment or dispersant spraying can be used. It is also very necessary particularly at sea to understand the fate of the oil once on the water and the ability of the oil to naturally dissipate, to move with wind and current, and thence to anticipate areas which might be threatened and thus to forestall such threats. At sea there are three natural phenomena which will affect the oil remaining on the surface. The first of these is evaporation which is quite rapid resulting in the loss from the water surface of hydrocarbons up to carbon numbers 10-12 usually within twenty-four hours depending on slick thickness, temperature and turbulence. For most North Sea crudes such a loss would be equivalent to fifty per cent of the volume of the crude oil.

Solution of certain lighter oil components can also occur, but is of little significance from the standpoint of oil removal, although such solution is significant ecologically. However the solubilities of such components are very low and particularly so when it is remembered that these components are preferentially soluble in the surface oil and are there very rapidly evaporated.

Natural dispersion is where minute oil droplets are formed by wave action and enter the upper water layer. These droplets, having a large surface area, will biodegrade in the water. The rate of oil loss from the surface due to this phenomena is significantly high and increased by wave action and thus sea state. Evidence to date indicates that in sea states around 2, with wave heights around 3 foot, such loss would be fifteen per cent in twelve hours. In some ways this effect compensates for the inability to utilise open sea recovery equipment in all but calmer seas and is now considered to have accounted for the apparently complete disappearance of some offshore spills following tanker incidents, particularly in heavy weather.

To assist governments and the oil industry in planning the necessary clean up capabilities, in ascertaining vulnerable coastal areas and during an incident to assess the degree of pollution threat, Shell and the Oil Industry Exploration & Production Forum have developed a computer model known as Sliktrak. This model, designed for application to an offshore spill, takes into consideration the two main oil-removing phenomena of evaporation and natural dispersion, whose effect can be assessed from readily available weather data. By then combining the oil removal rates which can include the effects of oil clean up at sea with oil drift due to wind and current, the oil's path across the sea can be predicted as can be the area at which it will hit coastline together with the quantity. This model, originally developed for the North Sea, has thus been used to develop numerous spill scenarios from different oil field locations, different well flow rates, to give overall or individual plots of oil spill trajectories and thus the potential impact at coastline. It also incorporates the ability to predict cost of clean up and to give an economic estimate of potential shoreline effects and clean up.

As in all incidents giving rise to major pollution threats in E&P operations, it is by effective attention to the areas of prevention, both in equipment and procedure, contingency planning and by understanding the extent of the threat that such pollution can be minimised.

OPERATIONAL POLLUTION

So far a great deal of attention has been given to the causes and methods of preventing, coping with and understanding the threat of oil pollution arising from E&P operations, the possibility of ongoing operational pollution which has also to be curbed effectively. Here, however, we are not looking at as complex a picture as might be the case in oil refining operations. The method and extent of treatment does vary, depending on whether one is considering offshore or onshore operations and on the potential environmental impact of the resultant discharges. In all cases the sources of such operational pollution can be categorised relatively easily. In both exploration and production operations one can expect oily effluents due to minor mechanical defects in the system, such as leaking pump seals, aqueous condensate and the necessity in certain cases to depressure vessels involved in the treatment of the product. In exploration operations two specific sources of oil pollution have to be considered, firstly where oil-based drilling mud is being used and secondly during well testing. In production operations the liquid flowing from the well is usually a combination of oil and water which has to be separated and the water disposed of.

In offshore operations the extent of necessary treatment will, to a large degree, depend on location, water depth and current dispersion and the effect of the treated discharge on potentially vulnerable species. Generally speaking water run off from platform deck areas, etc. will not be greatly contaminated with oil leaking onto the platform, but being a highly variable flow, it cannot be combined with produced water treatment. A typical approach is to pass this water to a large diameter deep caisson whose outlet is well below the water depth. This provides good residence time for gravity separation to occur and the separated oil would then be removed from the water surface in the caisson by pumps. The oil-free water then flows into the sea from the bottom of the casing.

Produced water is liable to be more contaminated with oil and undergoes treatment prior to discharge to a similar caisson. Such treatment is usually based upon two approaches or a combination thereof. Gravity treatment using separators similar to the corrugated plate interceptors used in refineries is one approach, often using banks of two in series. The second approach is the use of flotation cells with several in series. Here oil droplets are removed to the water surface by the use of finely divided air sparged into the bottom of each cell. Often a gravity separator is used upstream of floatation cells to remove bulk oil and to enhance droplet coalescence, assisted in many cases by chemical addition. As with most critical systems in offshore operations, separator trains are frequently duplicated.

In exploration, whether it be onshore or offshore, the use of oil-based drilling mud required special treatment. Oil-based drilling mud is not often used being an expensive alternative to water-based mud; certain rock formations can, however, be adversely affected by the presence of water, which can, for example, cause swelling and weakening, and here the use of oil-based mud can be necessary. The mud leaving the well is treated to remove rock cuttings by graded shakers and sieves to enable its re-use. Potential pollution problems arise through two sources, firstly the dispersal of oily cuttings and secondly mud spillage during the drilling operation when the drill pipe is removed from the well. In the latter case such spillage is almost invariably routed back to a storage tank for treatment and re-use. Oily cuttings offshore are extensively washed to remove oil prior to dumping. The oily washings then are routed to the oily water caisson. Onshore this problem is less critical since the oil in the cuttings will biodegrade on exposure to air. Spent drilling mud is usually disposed of by incineration.

Well testing when oil is flowed from a well to test the reservoir potential is nowadays carried out through a flare, or perhaps onshore into specially bunded areas from which it can be removed.

Onshore treatment of oily water produced from the reservoir is far more akin to the treatment of refinery effluent and will invariably include gravity separation followed by further treatment depending on the capability of the receiving waters to tolerate the biological load. In some cases both onshore and offshore it may be practical to reinject this water into the reservoir which of course obviates the need for treatment. This however will depend on the characteristics of the reservoir since in some cases some reinjection can inhibit the recovery of oil from the reservoir.

* * *

To conlude, the operational oil pollution problems in E&P operations are probably easier to handle than in, say, oil refining, and the major concern must be the potential, albeit statistically small, threat posed by a blowing well. To provide a complete picture of this latter problem it is necessary to understand the techniques involved in E&P operations to in turn appreciate the preventive and remedial measures undertaken, it is hoped, however that this brief account has given some idea of the problems involved.

5
Accidental spills from tankers and other vessels

W.O. Gray, BE, BSE
Logistics Dept. Exxon Corp

The purpose of the present chapter is to give an overview of seagoing oil tankers, types of accidents, and oil spillage from tanker accidents. Ship design features adopted for greater safety and environmental protection are described as well as operational measures for improving the safety of shipping of all types. Finally the chapter touches briefly on possible accidental oil pollution from ships other than oil tankers as all powered vessels using oil as fuel are potentially a source of oil discharge to the seas.

The increase in the use of petroleum and the growth in size and numbers of tankers were discussed in chapter 1. Figure 7 shows world oil movements by sea. There was an extremely steep growth curve until 1973, at which time a sudden reversal in trend took place. This was caused by the oil embargo of late 1973 which was accompanied by a quadrupling of oil prices and a dramatic decrease in growth of oil consumption and accordingly at-sea oil movements. In fact, the total amount of petroleum moved at sea dropped for a number of years and it was not until 1977 that shipments were equal to the level they had reached in 1973. By that time total demand for tonnage to transport oil at sea was of the order of 30 per cent less than had been anticipated just a few years before. Consequently, starting with 1973, a dramatic of tanker tonnage began to develop. It is with us now and is expected to last many years into the future.

With the radical reduction in demand for tanker tonnage which took place a few years ago it is clear that major tanker new building programmes will not be undertaken for several years in the future. Accordingly, efforts to minimise or reduce accidental oil pollution should be centred upon the operation of existing tankers and the safety features inherent in these vessels as they are currently constructed. Secondly, because of the prolonged surplus of tonnage projected, the tanker shipping community is obviously in a severely depressed state. This in turn causes valid concern with regard to the maintenance and repair policies of tanker owners which will need close watching and unusual diligence to assure that the quality of the tonnage is maintained so as not to pose undue safety or environmental risks.

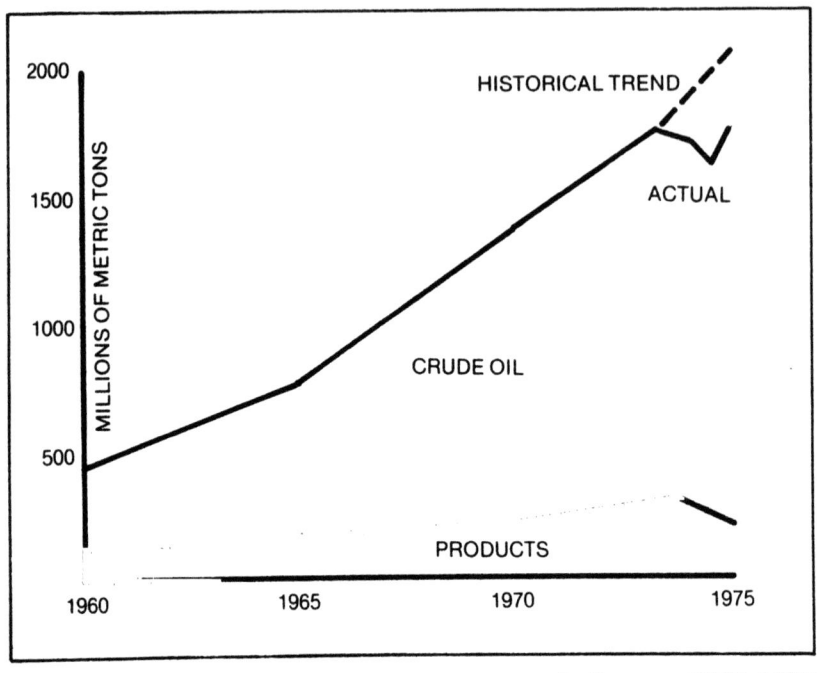

Figure 7: *World movement of oil at sea 1960-1975*

Figure 8: *Comparison of sizes of 250,000 DWT and 400,000 DWT tankers.*

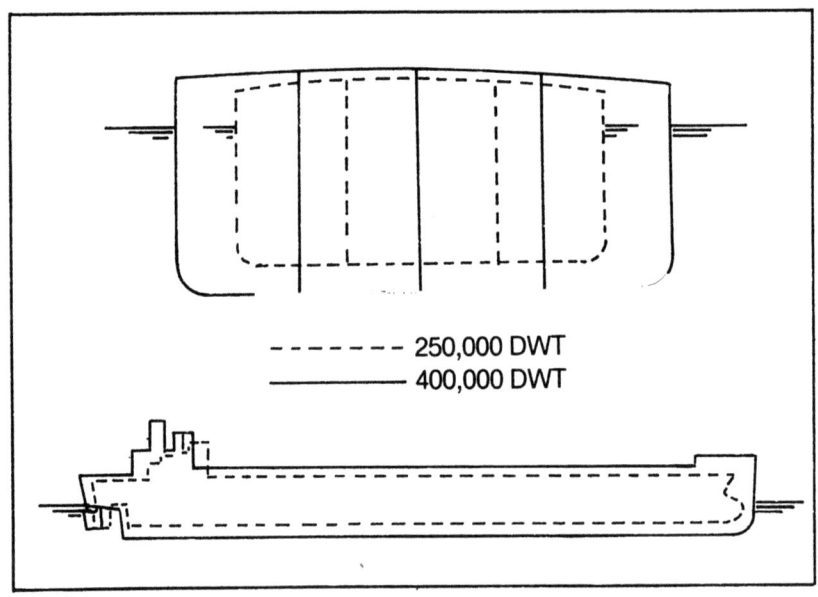

TANKER CHARACTERISTICS

The basic design of most modern oil tankers is basically very similar. It is described here in terms of typical very large crude carriers (VLCCs) or ultra large crude carriers (ULCCs), the type of tankers of greatest interest generally when discussing accidental pollution. In figure 8 cross-sectional and side views of tankers of 250,000 and 400,000 deadweight tons are shown. It can be seen that tankers are long relatively shallow vessels with a rectangular cross section. This boxy cross section is maintained throughout much of the ship length tapering to ship shaped form at the bow and the stern to provide seaworthiness, manoeuvrability and efficient propulsion. The rectangular cross section has a number of advantages. First, it allows the maximum volume of cargo to be carried on the minimum draught thereby efficiently utilising the depth of water in which the ship will operate. Second, a rectangular cross section provides for economical construction, and finally a closed plate box girder, which is the way a structural engineer would describe a tanker hull, represents the most efficient and reliable possible type of structural beam. In order to achieve maximum carrying capacity, it is desirable to minimise the hull structural weight consistent with reliability requirements. Accordingly, VLCCs are typically constructed of medium and high strength steels with a maximum thickness on the order of 1- 3/8 inches, a dimension roughly equivalent in relation to the ship's overall dimensions to the thickness of the paper wrapper on a loaf of bread. Quite obviously, such a thin structure would be totally unable to resist the loads to which it is exposed were it not for substantial internal strengthening. Figure 9 indicates the primary structural strengthening members and partitions, known as bulkheads. These plate structural members divide the ship into a series

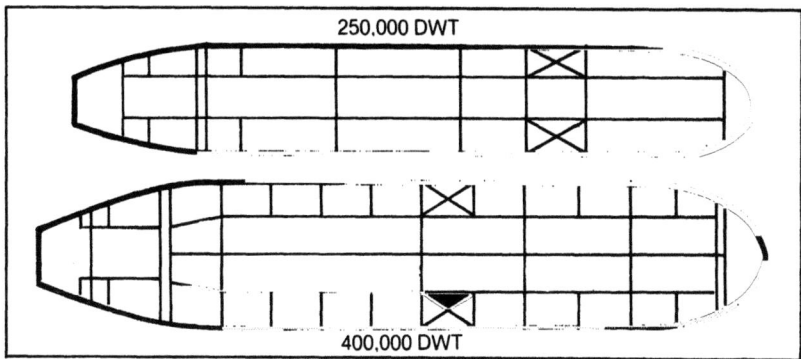

Figure 9: *Primary structural strenghening members and partitions in 250,000 DWT and 400,000 DWT tankers.*

of tanks or compartments each water-tight or oil-tight from each other. They also provide the strength needed to assure sufficient rigidity of the main hull to withstand both wave loads and internal loads from the weight of cargo and the ship itself. Propulsion in VLCCs is usually by means of two steam turbines geared to a single propeller with steam normally supplied by a single main boiler. There is also a small auxillary or back-up boiler. Steam is also the source of power for driving the turbines which turn the ship's four centrifugal cargo pumps needed to unload cargo and ballast from the ship.

As an alternative to steam turbine propulsion, a number of VLCCs and the majority of new smaller product tankers commonly employ large diesel engines for main propulsion with the diesel engine most commonly being directly connected to the propeller. In this latter type of design it is aslo common to fit a boiler as well to supply steam for the cargo pump turbines.

The efficiency of design and construction of modern tankers can be evidenced in several ways. First, the overall weight of the whole structure, machinery and accommodations of a typical VLCC represents only about 12 or 13 per cent of the total loaded weight or displacement of the ship. The remaining 87-88 per cent is devoted mostly to productive cargo carrying capacity and, of course, to fuel and provisions for the ship and crew. By contrast, in tankers constructed shortly after Second World War in the 20-30,000 ton size, as much as 25 per cent or more of the total loaded displacement would be accounted for by the total weight of the ship.

Another substantial measure of the efficiency of the larger crude tankers is their propulsive efficiency. At the fully loaded speed of 15 knots, which is common for tankers over a range of sizes from 10,000 to 500,000 tons and more, very large decreases in fuel consumption can be noted. For example, on a typical 11,000 mile loaded passage from the Persian Gulf to the English Channel, a ship of 20,000 tons deadweight will require about 900 tons of fuel and it will deliver about 23 tons of cargo for each ton of fuel consumed. The VLCC of 250,000 tons will consume 4,800 tons of fuel, but it will deliver almost 250,000 tons of cargo, or over 50 tons of cargo for each ton of fuel consumed. This is double the efficiency of the 20,000 ton tanker. In the latest ships of the 500,000 tons class, fuel consumption is 6,400 tons for the 11,000 mile voyage, but this equates to almost 80 tons of cargo delivered for each ton of fuel consumed. In addition, because of the smaller amount of engine horsepower required relative to the size of ship and only modest changes in crew accommodation requirements as the size of the ships increases, construction cost for vessels of larger size does not increase nearly in proportion to the increase in vessel size.

Figure 10 shows a rough indication, for an 11,000 mile voyage, of the cost of transporting oil by tanker comparing a 50,000 ton tanker typical of the 1950s with the 250,000 tonners which become the workhorse of the early 1970s. It can be seen that transportation costs have decreased by more than fifty per cent by going to the larger vessels. In addition to economic incentives there are substantial safety benefits from using larger ships. A much smaller number of tankers is needed to carry current quantities of oil than would otherwise have been the case if tanker size had not increased as it did. With the requirement for oil movements at sea nearly quadrupling in a little over twelve years it is doubtful whether this tremendous increase in transportation capacity could have been accomplished without the major increases in ship size. It is questionable whether it would have been possible to find, recruit and train the number of officers and crews needed to man the increased number of tankers which would have been required if ship size had stayed the same.

Figure 10: *Cost in dollars per barrel of transporting oil by tanker over a typical 11,000 mile voyage.*

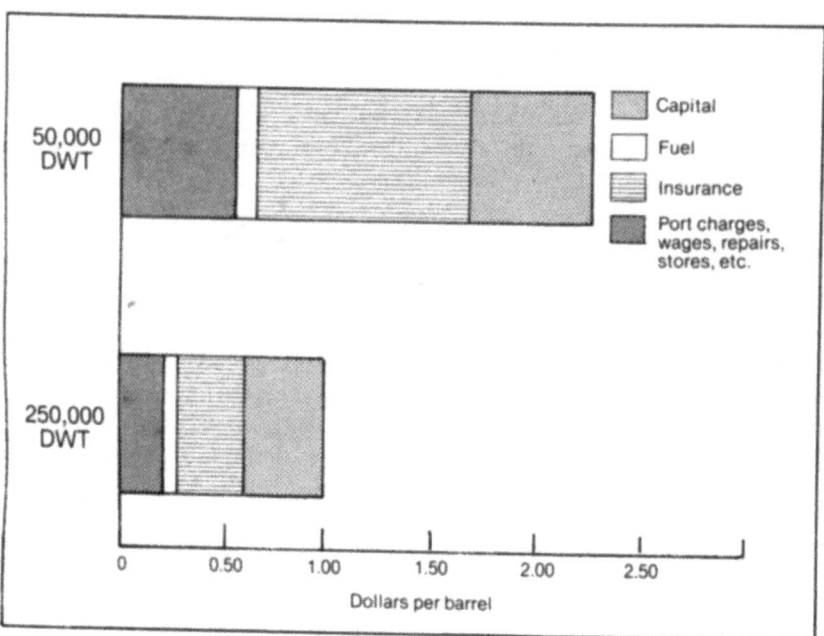

With regard to crew size, for tankers anywhere in the size range of 15,000 to 500,000 tons a crew of 30 has now become common. By contrast with older tankers in which nearly all tasks were performed manually, it is common in modern tankers, both large and small, to arrange for operation of the main engine entirely from the bridge by a watchstanding officer. Similarly, control of all valves and pumps to load and discharge cargo is often accomplished from one central cargo control room. The tankers of today are filled with modern technology in their engine rooms, on their bridges and in the cargo systems. It requires well-trained and skilled personnel to operate and maintain these ships.

In addition to the need for large tankers to transport crude oil there is the continuing need for smaller tankers to transport refined petroleum products and fuel oils. In product trades, not only are there demands for small individual parcels of cargo, but the destinations for these cargoes are far more diverse than for VLCCs which bring their crude oil to large refining centres. In product service there are literally thousands of terminals or harbours to which these ships must trade. Accordingly ships of modest size, generally in the neighbourhood of 20-60,000 DWT are most suitable. Product tankers, are fundamentally similar in basic design to the VLCCs. The main difference to be borne in mind is that ships in product service commonly make much shorter voyages with many more port calls per year than large crude tankers. This causes the smaller tankers to be in harbours and coastal waters for a much greater percentage of their operating life than is the case for VLCCs. Therefore to the extent that the harbour and open sea environments pose different risks and dangers to the tankers this factor should be borne in mind in looking at the accident record for the two basic classes of tankers.

ACCIDENT RECORDS

Accident data for ships of all types have been accumulated for many years. There are probably as many ways of displaying and interpreting these data as there are sources of data. None, however, give an absolutely precise picture of those aspects of ship operation which are most dangerous or most in need of correction. Nonetheless a general understanding of the accident data can be extremely helpful in drawing general conclusions or guidelines for future improvements.

Figure 11 shows the historical accident pattern over the last century for merchant shipping of all types. It clearly reveals that as merchant shipping moved from the days of sail at the turn of the century into the days of the steam and diesel motor ships, there was a steady decline in the annual tonnage lost as a percentage of tonnage at risk. In 1960 this trend suddenly was reversed and the record deteriorated for five years after which it once again resumed a pattern of improvement which has been sustained into the decade of the 1970s. While one can only speculate on the reasons for the deterioration in performance during the 1960s, two possible explanations suggest themselves. First, a substantial number of merchant ships built during the war were still operating at that time. By the 1960s they were quite old and one naturally questions their state of repair and maintenance. Secondly due to the tremendous growth of world commerce, as typified by the growth in oil movements at sea already described, there was a large entry of new firms and ship operators into the maritime business. Both the tremendous growth rate and the entry of new firms into the field are factors which could have contributed to the deterioration in the safety record which clearly

Figure 11: *Annual steam and motor vessel tonnage lost as percentage of total tonnage at risk.*

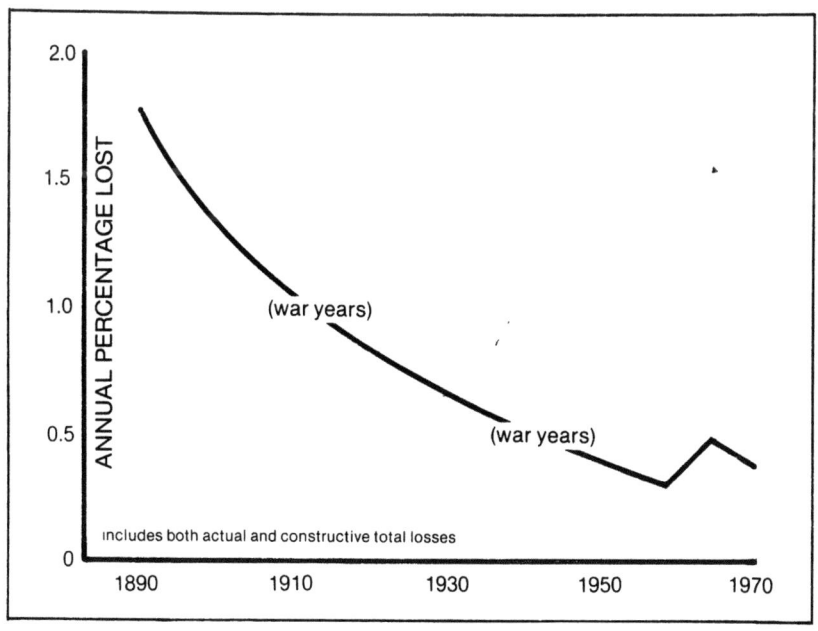

took place. The data in figure 12 present the average loss ratios for tankers and non-tankers over a series of five year periods in the period following the Second World War. Two patterns are clearly evident: firstly, over the period in question, tankers have had a decidedly better total loss record than have non-tankers; secondly, during the period of 1959 to 1973 there was a marked deterioration in the total loss record for tankers whereas a similar trend is not evident for non-tanker shipping. A further examination of these figures has shown no clear trend to explain this deterioration.

In table 5 the serious casualty data are separated into four size classes to determine whether or not significant trends by size class exist. Here again distinct trends are difficult to discern with two possible exceptions. First, the largest size category of tankers, those of 150,000 DWT and above appear to have a slightly lower rate of serious casualties than do the other classes below 150,000 DWT. Secondly, the early years of the larger ships of 150,000 DWT and above showed a record inferior to that of the smaller ships. While the rather poor record in the early years of the larger ships is attributable to a small number of serious casualties in a brand new class of shipping, the reasons for the other trends are not obvious. Factors to be borne in mind are as follows. First, as already mentioned, smaller tankers generally trade more frequent-

Figure 12: *Annual average loss ratios of total tonnage at risk by vessel type.*

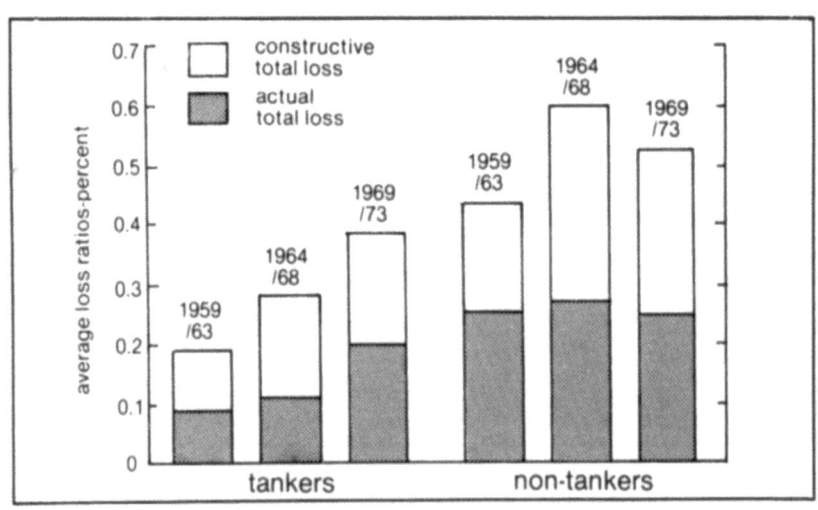

ly in near coastal and harbour areas and are more frequently involved in transfer of cargo. As these conditions represent greater exposure to certain hazards one would expect a somewhat higher incidence of serious casualties whereas none is clearly evident. Secondly, generally speaking the age of the smaller ships as a group is greater than that of the larger ships, and here again one generally expects to find a somewhat greater rate of serious casualties with the older ships than with newer tonnage. Aside from these general observations, however, it is not believed that the data in table 5 show any remarkable trends.

Although not directly related to environmental protection, an indicator of great interest to ship operators is personnel casualties and lives lost. The data in table 6a show the causes and numbers of lives lost on tankers over a nine-year period. The average figure of 116 deaths per year can be put in perspective by reference to the fact that at any given time, there may be approximately 4,000 tankers at sea each with an average crew in the neighbourhood of 30-40 seamen. Accordingly, between 100,000-150,000 seamen may be on tankers, and it is this figure to which the average of 116 deaths per year should be compared. Other studies indicate that merchant seafaring aboard tankers ranks generally with heavy industry as regards personnel accident hazards.

Table 5: *Serious casualties to tankers of 10,000 DWT and more by year and size of ship (1968-1976).*

	10,00-24,999 DWT		25,000-44,999 DWT		45,00-149,000 DWT		150,000 DWT and above	
	Serious Casualties	Rate per 100	Serious Casualties	Rate per 100	Serious Casualties	Rate per 100	Serious Casualties	Rate per 100
1968	47	3.38	17	2.26	15	1.91	0	–
1969	40	3.01	15	1.94	17	2.04	4	7.02
1970	33	2.60	8	1.02	16	1.83	3	2.48
1971	20	1.61	21	2.57	15	1.67	8	4.04
1972	32	2.63	18	2.21	17	1.85	7	2.62
1973	24	2.08	19	2.25	20	2.14	2	0.57
1974	23	2.04	19	2.17	22	2.24	5	1.10
1975	27	2.59	23	2.64	28	2.65	10	1.72
1976	23	2.41	29	3.49	30	2.74	11	1.58
Total	269	2.51	169	2.30	180	2.15	50	1.82

Source: *Tanker Casualty Sub-Group.*

Table 6: *Lives lost due to serious casualties to oil/chemical tankers 1968-1976.*

a: *by year*

	number of lives lost
1968	98
1969	124
1970	88
1971	127
1972	220
1973	65
1974	84
1975	115
1976	119
Total	1040
Average per annum	116

b: *by category of casualty*

Cause	number	%
Cargo tank fire/explosion	227	27
Collision*	203	19
Ship sinking	195	19
Men working in tanks	109	10
Missing ships (no known or suspected cause)	107	10
Engine room fires/explosions	47	4
Other fires/explosions	26	3
Deck fires/explosions	23	2
Machinery troubles	19	2
Pumproom fires/explosions	18	2
Other causes	18	2
Total	1040	100

*fire and explosion following collision caused 190 out of 203 lives lost due to collisions.

The data in table 6b show the dramatic figure of no less than 45 per cent lives lost were a result of fires or explosion. Considering the volatile cargoes carried by tankers, this statistic is not surprising. Nonetheless it underlines the continuing concern which ship owners and crews share in fire prevention measures and firefighting means aboard tankers.

In chapter 1 mention was made of a survey of tanker casualties in which oil was spilt over the period 1969-1973. It was noted that collision, grounding and structural failure accounted for over 75 per cent of all oil spilt; but considering the numbers involved there were relatively few large accidents. When these were grouped by sizes of tankers causing oil outflow (as has been done in figure 13) it shows rather clearly that in terms of oil outflow, or oil spilled per ton of carrying capacity, the larger and newer ships clearly have had a superior record. Again it should be noted with caution, however, that the smaller vessels generally have the greater number of port calls and are trading in congested areas more frequently than the larger ships, and that the majority of the oil outflow comes from a relatively very small number of incidents and, accordingly any of these trends can be dramatically influenced by a single accident in a given year.

DESIGN SAFETY FEATURES

Having looked at the types of accidents in which tankers can be involved, it is of value to look more closely at the basic design features of a typical tanker to see how these contribute to protection of the tanker, its crew, and its cargo. Firstly, every seagoing vessel of whatever type is assigned a load line or deepest

Figure 13: *Annual oil outflow (in tons of oil outflow per 1000 tons* DWT *) by tanker size — all categories 1969-1973.*

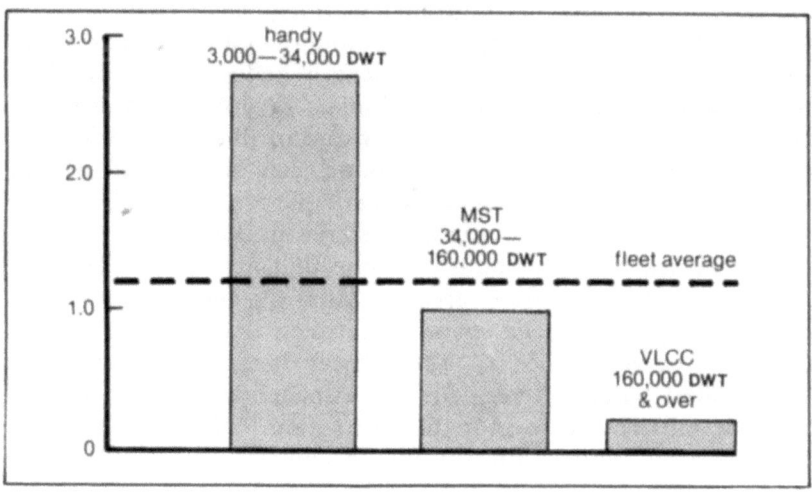

draught to which it may safely be loaded. The deepest load line is marked at the sides of the ship by the familiar Plimsoll mark. There are three fundamental concepts behind the assignment of a load line to any ship. First, it is to assure sufficient freeboard, or distance above the water to the deck of the ship to provide a safe working platform for its crew. In reality, on modern tankers, this concept has little remaining significance as the crew of the ship seldom, if ever, works on deck while the ship is laden. Secondly, the value assigned to the load line is intended to assure that the freeboard is sufficient to eliminate undue risk of heavy seas breaking through hatches on the main deck of the ship. In the case of tankers with liquid cargo, the hatches are in fact very small and accordingly this danger is virtually non-existent. These two factors are reflected in the permissibly deep draught and low freeboard assigned to tankers. Third, the load line of the ship should be assigned to assure that the loaded ship will have sufficient additional or reserve buoyancy so that in the event of damage or rupture to empty compartments while the ship is loaded, the ship will not sink farther than to a point where she will remain afloat with the main deck still above water. Application of these various principles to the typical modern tanker produces a load line nearly eighty per cent of the distance from the keel to the deck.

The second fundamental safety factor inherent in the design of the ship is the internal compartmentation or subdivision of the ship's total internal volume into a large number of small compartments. This can clearly be seen in figure 9 comparing 250,000 and 400,000 DWT tanker designs. In addition providing small individual compartments for containing parts of the cargo, a number of additional features can be seen. At the bow, there is always a bulkhead running completely across the ship at least 5 per cent of the length behind the bow. This bulkhead is known as the collision bulkhead, and no flammable or dangerous cargoes may be carried forward of this point. Additionally, no accommodation for personnel can be forward of the collision bulkhead. The reasons perhaps are obvious, but it provides protection to the ship and its crew in the event of striking another vessel. Next it should be noted that near the aft end of a tanker the cargo spaces are separated from the engine spaces by at least two bulkheads. The space in between is either known as a cofferdam or void space, or alternatively it may be occupied by water ballast which is emptied in the loaded tanker, or in some cases fuel oil. The reason for this protection is that in the engine spaces of the ships with various machinery components, the presence of sparking sources which may cause ignition will always exist. Accordingly, two bulkhead separation of potential ignition sources from the cargo spaces containing volatile cargo on tankers is required by regulation.

The other significant safety feature from compartmentation in tankers is that in the event of rupture of the tanks there are many tanks which will inevitably remain intact and therefore contribute to survival capability and stability of the damaged vessel. In this respect, because tankers commonly contain two or more longitudinal bulkheads running the length of the ship, the extent of tank damage in an accident is generally a great deal less than in other types of ships where cargo compartments commonly extend the full breadth of the ship from one side to the other. In reality, if a loaded tanker is ruptured in the bottom, instead of water running into the ship, a small amount of cargo drains out so that the ship lifts slightly out of the water rather than sinking from flooding. This is shown in figure 14.

The most important single safety feature for the ship is its basic hull structural strength. This must be controlled to adequate standards in order to prevent complete or catastrophic failure of

Figure 14: *Stability and buoyancy in groundings — single bottom tankers.*

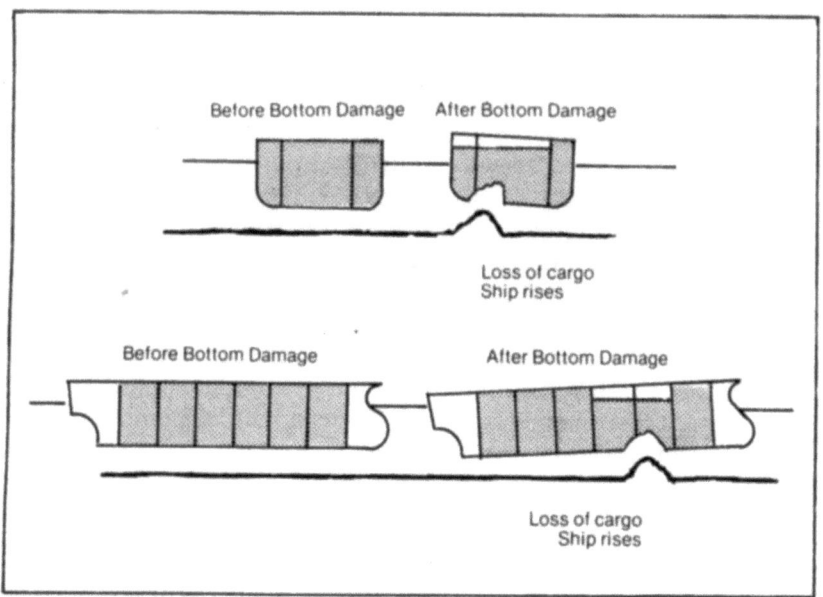

the hull. A superficial comparison of the hull structure of a modern VLCC with that of a typical tanker of the 1950s will show no apparent differences. In reality there is a vast difference due to the technological changes which have taken place in the interval. First, the emergence of high speed electronic computers allowed a sophistication of structural analysis and design techniques which had not previously been possible. The techniques of three dimensional frame analysis and finite element stress prediction used in the aircraft industry and the ship structural design field are virtually identical. The same degree of sophistication has been found desirable both from the point of view of structural reliability and design hull weight deficiency. Secondly, the ability to measure and then analyse the complex ocean wave forces applied to a ship's hull girder was developed from a highly theoretical exercise to practical reality about this time. This also would have been totally impossible without the ability afforded by high speed electronic computers. Finally, due to the high demand for tankers of greater carrying capacity, structural engineers had little past experience to guide them. Accordingly they had to adapt new and more sophisticated techniques to assume reliability of the ship's hull. During the same period there has been a completely new approach to the method of building ships. These modern ship-building techniques have resulted in a much higher quality product, both in terms of the materials used and of welding fabrication. Now sufficient experience has been gained to confirm the superiority of both the basic design and the hull construction procedures for virtually all the large tankers built during the 1960s and early 1970s. These features and the improved engine reliability form the basic safety features built into tankers.

Contrary to popular opinion, requirements for safety features are not left solely to the individual preference or different tanker owners. Most of the features in regard to hull structural strength and machinery reliability are the subject of classification rules and regulations to which ships must be constructed, and maintained, in order to obtain insurance for the ship and its cargo. There are several classification societies around the world whose sole job it is to make improvements to the structural reliability criteria for the basic hull and machinery components of ships of all types. Accordingly responsibility rests with a combination of classification societies, ship builders, and tanker owners, many of whom add additional design features of their own preference to increase the safety of the ship and its crew. Some of these will be discussed in a moment.

Some of the most significant basic safety features such as load lines, stability criteria, and personnel safety and fire protection are the subject of internationally agreed conventions passed to provide uniform international standards dealing with the fundamental safety elements of shipping. An early example of this approach dealt with lifeboat capacity and emergency radio procedures stemming from the lessons of the *Titanic* disaster in 1911. The rules are now included in the so-called SOLAS (Safety Of Life At Sea) Convention.

More recently, a series of new design requirements and regulations dealing principally with protection of the marine environment in order to minimise oil outflow, has been produced by IMCO.

The first substantial move in this direction dealing with minimising oil outflow from tanker casualties, was the so-called hypothetical oil outflow or 'tank size' regulations, first accepted in 1971 as amendments to the 1954 IMCO Convention. The purpose of these regulations is to limit the size of a tanker's cargo compartments to reasonable values thereby minimising the potential for oil outflow in the event of either a grounding or collision accident. While admittedly hypothetical in nature, they do provide uniform and consistent standards for ship designers throughout the world to assess the relative pollution potential of new ship designs from grounding and collision accidents. The 'tank size' regulations as they are familiarly known actually provide three criteria, all of which must be satisfied. First an absolute limit on the length of any given cargo tank is established, and secondly, absolute size limitations on cargo tanks are established as well. Because of the form of regulations, the first two criteria, tank length and tank size, tend to control the compartmentation of smaller tankers ranging up to approximately 200,000 DWT in size. The third criterion 'hypothetical oil outflow', tends to set the tank arrangement for ships in the VLCC and ULCC classes. The overall result of the 'tank size' regulations has been to increase somewhat the number of cargo tanks with which ships are built, particularly for ships of 400-500,000 tons.

One of the features debated at great length during IMCO's development of its tank size regulations was whether or not to require double bottom, or double hull tankers. After considerable study of accident and ship design data, IMCO first concluded in 1971, and later reaffirmed at its 1973 and 1978 conferences, that it would not be wise to make ship designs with either double hull or double bottom construction mandatory. Despite the fact that these types of hull configurations continue to draw strong support from many in the public and environmental sectors, the reasoning of the world's tanker design experts has been that such construction might provide little or no additional protection, and could in fact cause increases in the amount of oil outflow in the event of accident circumstances. This has particularly been the belief of those who have opposed the double bottom construction, which although it seems continually to appeal to certain in the United States, has found no support from any other maritime nations of the world.

There are two basic reasons for not favouring the double bottom structure in a grounding accident. While unquestionably capable of limiting some of the initial oil outflow following a slow speed or a low energy grounding, it could also increase the difficulty of salvaging the stranded vessel in the event that outer bottom structure is ruptured. This is shown diagrammatically in figure 15 which should be compared to figure 14. With the rupturing of the outer skin and flooding of the empty double bottom there is a loss of bouyancy thus settling the tanker more firmly on the bottom. To free the ship the lost bouyancy will have to be

Figure 15: *Stability and buoyancy in groundings – double bottom tankers.*

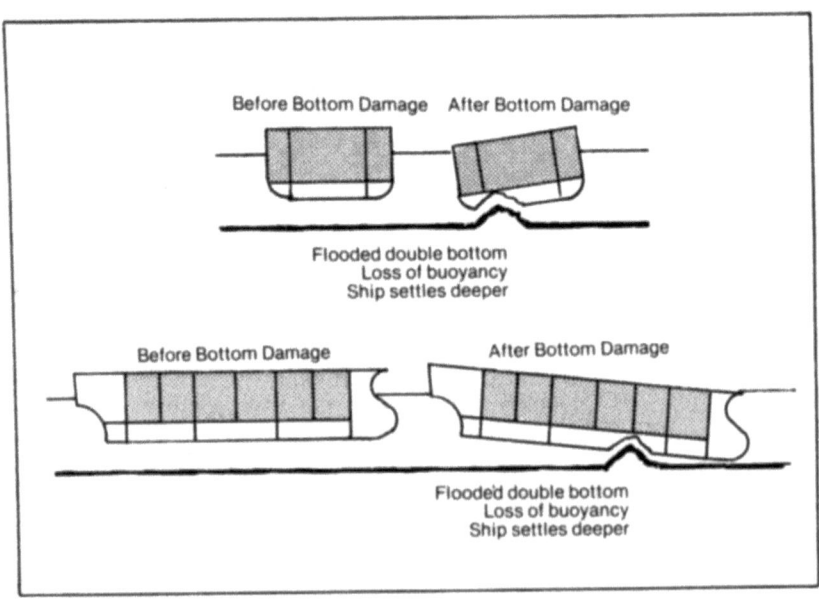

regained either by a greater lightening of cargo from the stranded ship or by jettisoning cargo over the side. By contrast the grounded single bottom tanker as previously explained will have risen slightly out of the water following bottom rupture. In the grounded single bottom tanker, ballast spaces, which must be equivalent in volume to those in the double bottom tanker, are available for controlling draught by buoyancy or flooding as long as they remain intact, whereas in the double bottom ship control over this bouyancy is automatically lost immediately the tank has been ruptured.

Oil outflow from grounding of loaded oil tankers on a worldwide basis has averaged approximately 40-50,000 tons each year. Two-thirds of that outflow or roughly 30,000 tons, is lost by tankers which become total losses as a result of their grounding, such as the *Argo Merchant* and the *Amoco Cadiz*. In these circumstances the double bottom construction could provide no additional protection to prevent the eventual breakup of the ship, and as just explained, it may in fact complicate the salvage operations increasing the likelihood of a bad accident becoming a total loss. Of the remaining third of oil outflow in accidents where the ship is not a total loss, it is problematical how much improvement could actually be made. Even the strongest advocates of the double bottom acknowledge that it will only be particially effective in these circumstances being capable of saving on a worldwide basis perhaps as much as 10-20,000 tons of oil outflow. It is this reasoning which is believed to have repeatedly caused the rejection of double bottom proposals at IMCO.

More recently IMCO has taken yet another step for new tankers in the endeavour to further minimise the amount, or possibility, of oil outflow in tanker accidents. This is the 1978 Tanker Safety and Pollution Convention protocols. Part of the IMCO response was a new concept to locate empty segregated ballast spaces in new tankers so as to gain an additional measure of protection. This so-called 'protective location' of segregated ballast had initially been developed in 1975 in the United States and put into regulations for ships in the USA in 1976. The fundamental basis of the 'protective location' concept is to provide a degree of protection by locating empty spaces at the side and or bottom shell of the ship so that the initial rupturing of the ship will not automatically penetrate cargo tanks, causing oil outflow and potentially fire and explosion, which so often occur in collision accidents. There are practical difficulties in application of the protective location principle as a total protection measure.

First, the amount of empty ballast tank volume available in a loaded tanker is not sufficient to provide a very wide or thick barrier around the loaded cargo tanks of the ship unless one is prepared to give up an unacceptably large amount of cargo carrying capacity which could only be compensated for by a large increase in the number of tankers used. Accordingly, the amount of ballast tankage to be located protectively was established as that amount of ballast needed for the segregated ballast concept for new tankers.

Secondly, even with the limited amount of ballast described above, which may amount to 15-20 per cent of the total internal volume of the ship, it is, of course, possible to provide a full double hull design in which areas immediately inboard of the side shell and bottom plates would be used only as ballast spaces. Arranging ballast in this fashion however would provide a very small distance between the shell of the ship and the outside boundary of the cargo tanks themselves. Accordingly the protection afforded would be more illusory than real in anything but the most minor grounding or collision accidents. As a result of this consideration and recognising the difficulties in grounding accidents from buoyancy problems, IMCO deliberations on protective location came to the conclusion that there probably was no single or unique best way in which to locate protective ballast spaces. Clearly a greater distance between the side shell and the cargo tanks is advantageous, but probably not in proportion to the increases in this distance particularly for grounding accidents. Secondly, it is problematical as to which type of accident, that is grounding or collision, is of greater importance and therefore deserves greater protective treatment. It was concluded based on accident records studied at IMCO that whereas the volume of oil outflow from collision and grounding accidents was rather similar over a range of years the impact upon personnel safety was drastically different. As table 6b demonstrated the number of tankermen deaths following collision accidents is very substantial, whereas death or injury to tanker crews seldom follow grounding accidents. Also there is only negligible potential for fire and subsequent explosion in groundings. Following this line of reasoning, many taking part in the IMCO deliberations were of the opinion that, to the extent that the protective location concept could afford an added degree of protection, they would prefer to see it applied to collision as opposed to the grounding accidents.

The flexible IMCO approach recognises that placing empty ballast tanks in new loaded tankers towards the sides and the bottom of the ship is probably most sensible. Figure 16 shows the types of location of segregated ballast tanks which will be encouraged by the protective location concept and some types of designs which will be discouraged. Finally in discussing this concept, it is important to record that those responsible for producing the protective location regulations clearly recognised that other measures for preventing grounding and collision accidents in the first place are of far greater importance than the protective location concept as such.

Two other design changes first adopted voluntarily by tanker owners and subsequently adopted as regulations at IMCO deserve mention. The shift from the familiar forward bridge design to a bridge aft concept was often regarded, when it was introduced in the 1960s, solely as a cost cutting or economy move by industry. While it is true that housing the entire crew at the aft end of the ship does produce economies in cost and slightly greater cargo capacity, the principal reason for adoption of this feature was to protect the crew from the possibility of fire and explosion accidents. With the midship house design familiar on older tankers, crew members living in quarters nearer the bow of the ship and navigating the ship from this point are both a great deal more vulnerable to the dangers of fire should the ship be in a collision, and secondly, the presence of men living and working over the cargo tank spaces which contain hydrocarbon gases causes a greater possibility for ignition. Sparking sources can come from machinery or simply the presence of people smoking. Accordingly, the bridge aft construction places the crew behind the most dangerous part of the ship and removes many of the possible sources of ignition from the tank space itself. This feature has now been required by IMCO regulations for new ships. As a side benefit, and much to the surprise of many in the industry, ship masters and pilots alike have almost uniformly commented upon the more advantageous position for conning or piloting the ship which the aft house provides. With the entire length of the ship in front of them it is far easier to observe behaviour of the ship while manoeuvring than it was with the more traditional forward bridge position.

Another notable safety feature of particular interest is inert gas systems the use of which is discussed in chapter 6. As is explained in the next chapter it is often necessary to wash cargo tanks during the ballast voyage when returning to a new loading port. During the tank washing process, using either water or crude oil, as is now increasingly being practised, high pressure jets of washing

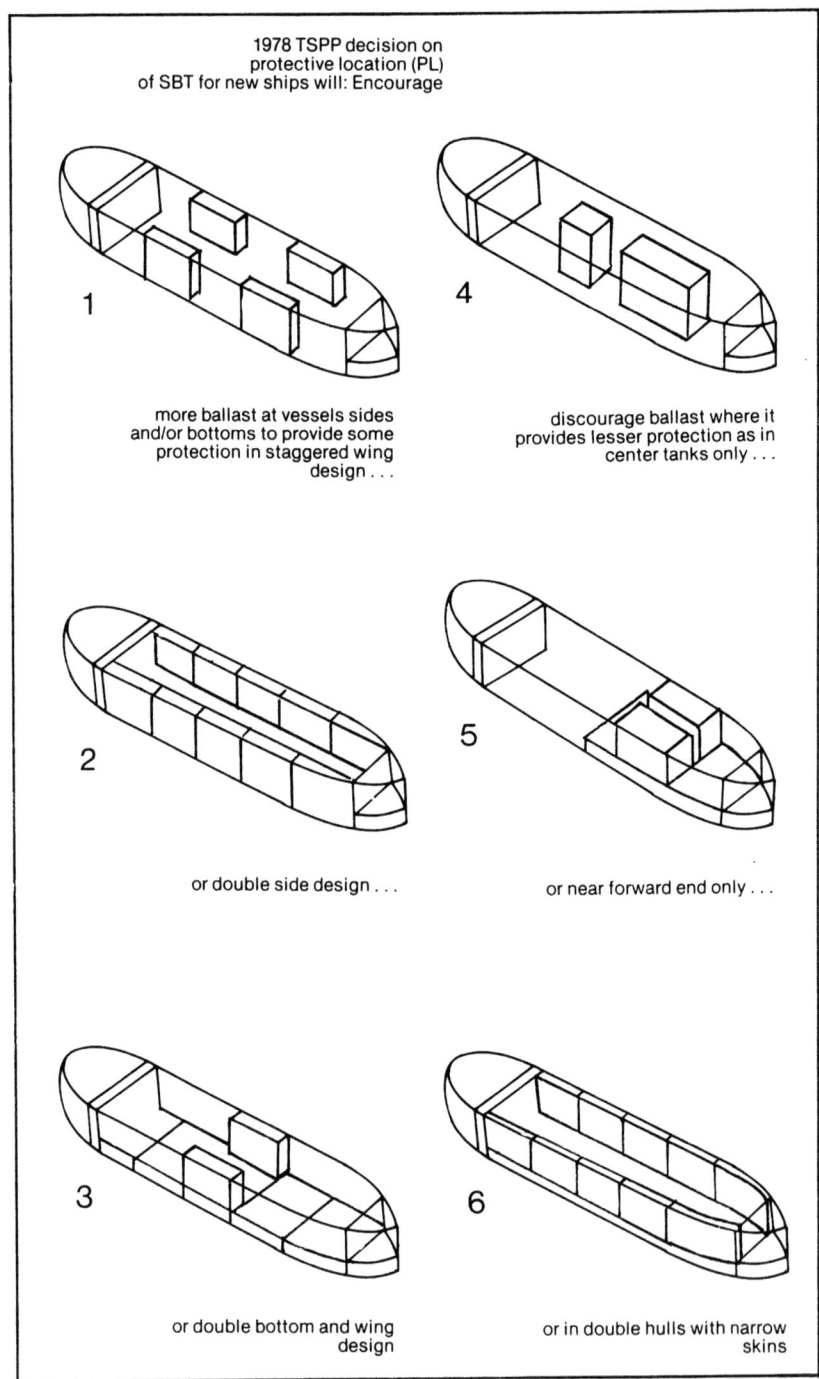

Figure 16: *Positioning of ballast in tankers.*

fluid are sprayed inside the tanks either by portable machines lowered into the tanks on hoses, or by fixed washing nozzles fitted in the tanks for this purpose. As a result large amounts of static electricity can be generated, sufficient, in fact, to cause incendiary sparks under certain conditions. This phenomenon has been known in the tanker business for years and operational safety precautions had been adopted to avoid either explosive mixtures in the tank or the probability of ignition sources. Nonetheless a series of explosions in 1969 involving three new VLCCs all conducting tank washing under similar circumstances convinced industry leaders that something more had to be done to minimise the possibility of tank fires and explosions. This is vital during tank washing operations, but also important during other operations such as cargo loading and discharging, and gas freeing of tanks or replacing the hydrocarbon and air mixture with air to permit work by men in tanks.

The solution adopted is inert gas which replaces the air in the cargo tanks and is incapable of sustaining combustion. Inert gas is most commonly taken directly from the exhaust or stack gases on tankers equipped with steam boilers; these exhaust gases are already themselves inert being made up of a combination of carbon monoxide, carbon dioxide and a large proportion nitrogen with very little oxygen. The quantity of inert gases available from the stack is more than sufficient to replace cargo in the tanks as the ship is being discharged. As the ship is being unloaded, the vapour space is continually filled with inert gas so that upon completion of discharge the ship contains a mixture of hydrocarbons and inert gas which will not support combustion even if an ignition source is present.

Figure 17 shows schematically the type of system needed to distribute inert gas from a ship's boilers to its cargo tanks. The gas leaving the stack must go through a scrubbing tower which both cools the hot stack gases and removes certain solids or a particulate matter by means of a seawater scrubber. The gas coming from the scrubber is forced by blower fans forward on the ship through a deck water seal. The deck water seal is a vital piece of equipment in the inert gas system as there must be a positive means of assuring that the flammable vapours of hydrocarbon and air from the cargo tanks cannot flow backwards through the inert gas distribution system into the ship's engineroom or boilers where explosion would certainly occur.

Starting in late 1970 and early 1971 a great many owners of larger crude oil tankers voluntarily took the step to refit inert gas equipment to many of their existing ships, and probably a majority of the crude tankers being built at that time were subsequently delivered with inert gas.

More recently at a conference called by IMCO in 1978, the requirements, which had initially been adopted in 1973 for inert gas to be mandatory in newbuilding tankers over 100,000 DWT, were extended to include not only new but many existing ships in considerably smaller size ranges. This decision was reached in the belief that, whereas the accident record for tankers had not clearly shown the need for inert gas either on rather small tankers, or on tankers with small capacity portable tank washing machines, there was little dispute that a substantial degree of additional protection could be achieved by the use of inert gas on tankers of any type. This type of reasoning had already lead a number of tanker owners to decide voluntarily to refit inert gas systems to many ships of their own fleets including smaller tankers of the product size.

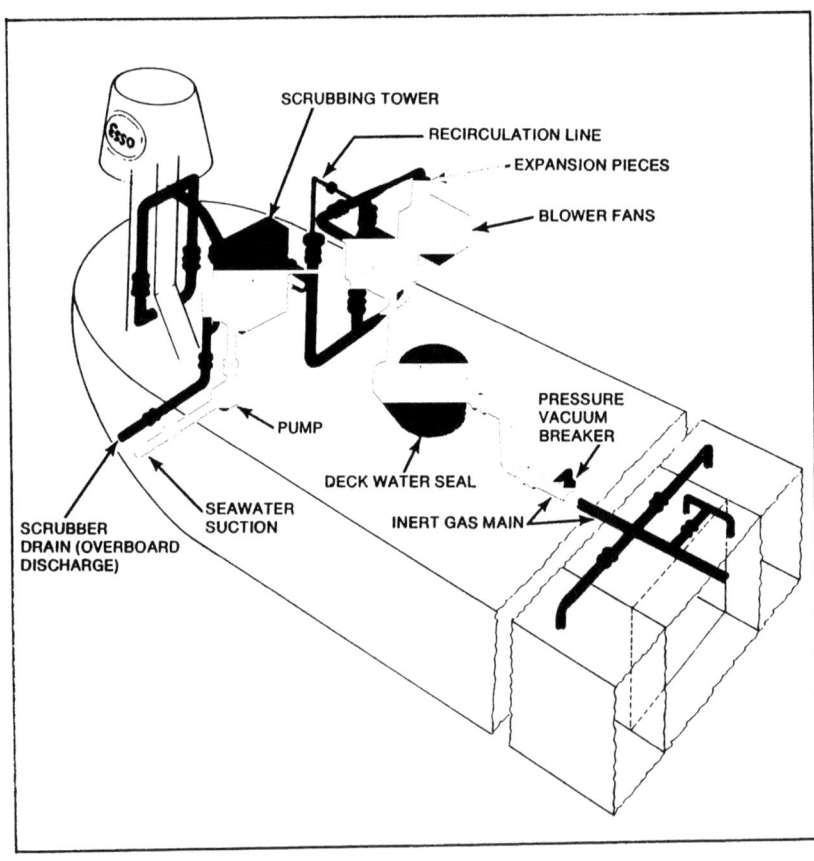

Figure 17: *Arrangement of inert gas distribution system.*

MANOEUVRING OF TANKERS

Discussion of the technical features of tankers up to this point has been concerned with reliability and safety on the high seas as well as in port. There are two other essential ingredients to safe tanker operation yet to be addressed. These include the manoeuvring characteristics of the vessels and finally the navigation of tankers and vessels of all types. Safe passage of a moving vehicle from one point to another, be it an aircraft, car, truck, or a ship, involves a direct interaction between man, (the pilot or driver) and the vehicle itself. This will be discussed in depth in chapter 12. In the context of the marine environment, the safe handling of vessels of all types is influenced by the following variables or factors. Firstly the man handling the vessel must be adequately trained and sufficiently skilled to control his command under all circumstances. This is without doubt the most critical single factor. The introduction in 1967 of shiphandling simulators is a major technical advance. They are valuable for giving training and practice to the shiphandlers, and for conducting manoeuvring research. This is true as well of model shiphandling facilities. An example of these types of facilities pioneered by Exxon since their early inception is shown in plate 11. Secondly the ship must be designed and maintained so as to be capable of

Plate 11: *Shiphandling simulation facility.*

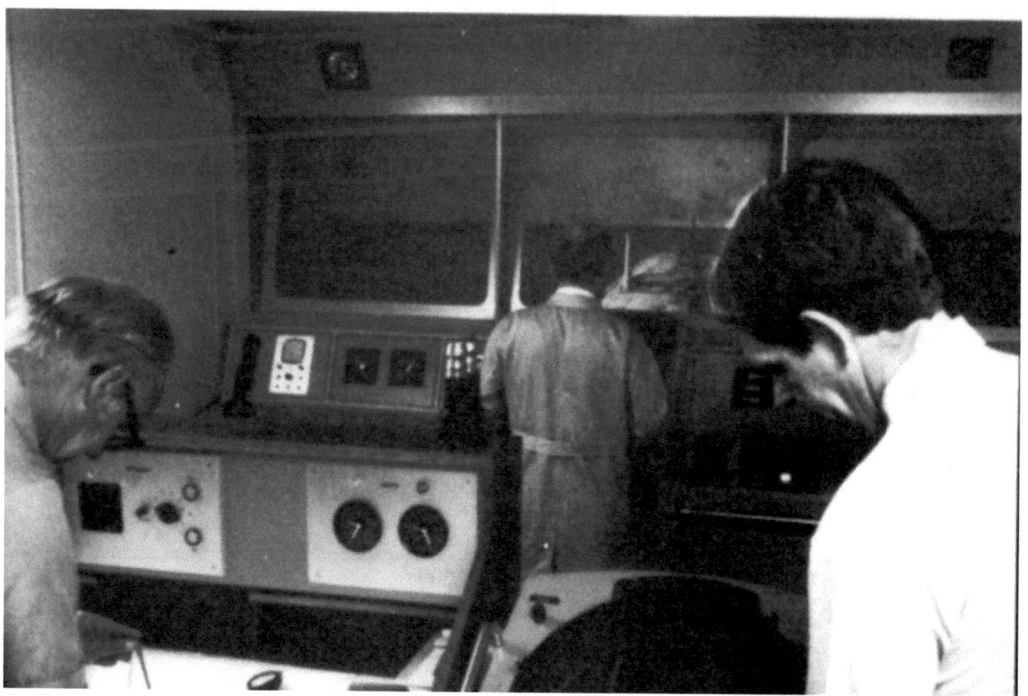

performing all necessary manoeuvres in an acceptable and predictable fashion; and the way in which the ship will behave must be known to its master. Thirdly the environment in which the ship is to be navigated must both be known to the master and suited to the ship's ability to manoeuvre. This concerns not only questions of current, wind and visibility, but also adequate space in which to turn and stop. Lastly there is traffic and navigational information which often constitute the principal environmental unknowns under randomly controlled conditions. Not only must all vessels have accurate position information continuously, but vessel movements and intentions must be clear to each other when in proximity of other traffic. Satisfying this requirement involves elements such as traffic separation, traffic control, bridge to bridge communications and procedural measures of this type.

With these fundamental criteria and definitions as background it is interesting to examine the inherent manoeuvring capability of tankers of various sizes. Those physical characteristics of the ship of greatest importance to its inherent manoeuvrability are its overall dimensions and its mass. Thus, it can be stated in a general way that although a 200,000 DWT tanker has a mass ten times that of a 20,000 ton tanker, because its length is only twice as great, the distance needed to turn the vessel will be approximately twice as great. As regards stopping of ships of various sizes, here again despite popular reports to the contrary, stopping distance of tankers of increasing size does not increase nearly in proportion to their increase in DWT. Fundamentally this is true for two reasons. First, the initial speed or full speed of the larger tankers is very little different from that of the smaller tankers. Secondly, the major factor contributing to the stopping of tankers or ships of whatever type from higher speeds is simply the resistance of the ship. The net effect is that the stopping distance of ships of the VLCC/ULCC class from full speed will be approximately three or

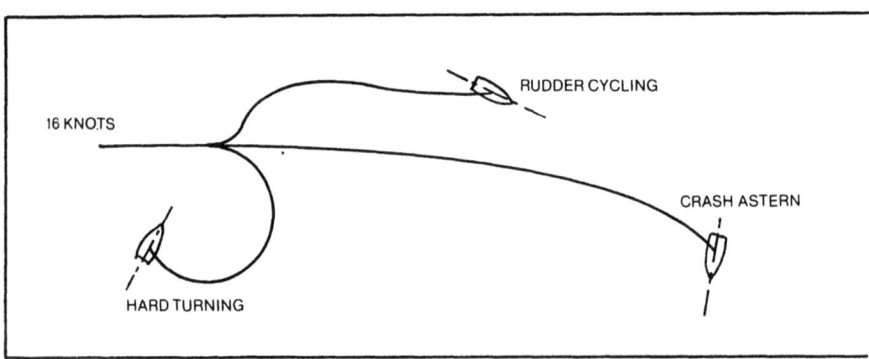

Figure 18: *Schematic comparison for emergency manoeuvres of large tankers.*

four times as great as that of ships 20-50,000 DWT.

Ships of most types are capable of turning a full circle in a distance approximately three times the length of the ship. Thus, a handy or medium size tanker of 200 metres length requires roughly 600 metres to reverse its course, a value which will be shown in a moment to be far less than the distance required to stop from high speed. A 500,000 DWT ULCC with a mass of ten times this size with an overall length on the order of 400 metres will by contrast require a distance of approximately 1200 metres or three times its length to make a full turn. It should be appreciated that, within these broad generalisations, different specific design measures such as size and rudder configuration, propeller design and other individual choices will have a noticeable but generally modest impact upon the basic manoeuvrability of the ship.

Far more important in the discussion of manoeuvrability is the matter of choosing the speed at which the ship is to be operated. Figure 18 will help to illustrate why this is so. Three evasive manoeuvres are available to the conning officer of a ship faced with an obstacle such as a navigational hazard or traffic requiring a course alteration. Firstly he can reverse his engine to full astern making the so-called crash-stop manoeuvre. Secondly he can put his rudder 'hard over' putting the ship into a tight turn, thus turning away from the danger and at the same time reducing the speed of his vessel as a result of the turn. His third choice is a manoeuvre developed by tanker officers in the early days of VLCC operation. This manoeuvre known as 'rudder cycling' involves several hard rudder manoeuvres together with slowing and eventual reversing of the engine in a series of linked manoeuvres or 'slalom' turns of the type a skier uses to check his speed while proceeding downhill. Figure 18 shows the path followed by a typical large tanker using each of these three manoeuvres from a

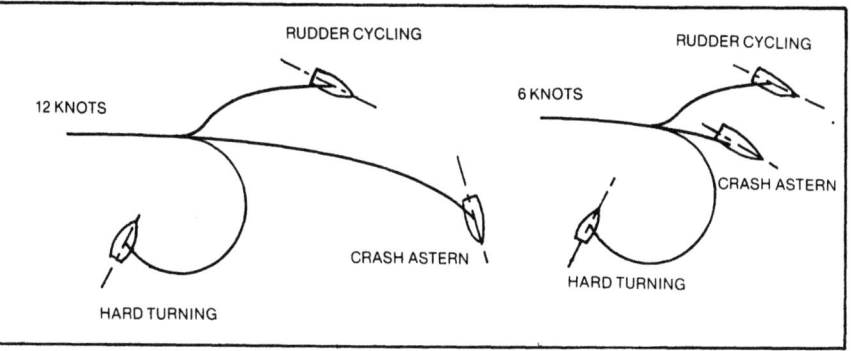

full speed of 16 knots, a more modest speed of 12 knots, and finally from a manoeuvring speed of approximately 6 knots. The space occupied by each of the manoeuvres may be judged in terms of distance travelled along the original course of the vessel as well as distance off course to either the left or the right.

It is apparent that, for a ship proceeding at full speed of 16 knots, the manoeuvre taking the least room is the hard turning manoeuvre. Crash astern from speeds of this type is a manoeuvre only used by shipbuilders and naval architects during ship trials as a relative measure of one ship compared to another. In reality, it has no place in the shiphandlers repertoire of manoeuvres in real circumstances. Even from a more modest speed of 12 knots, it can be seen that a hard turning manoeuvre uses a great deal less space than either of the other two manoeuvres. Nonetheless a substantial reduction in the distance travelled in both the crash astern and rudder cycling manoeuvre is evident. By contrast very little if any change can be detected in the path followed during the hard turning manoeuvre at various speeds.

The figure on the right from a manoeuvring speed of 6 knots, which is typical of a value chosen during the latter stages of harbour approach, indicates that any of the three manoeuvres can bring the ship to a stop or evade obstacles with a small amount of space being required for the manoeuvre. Interestingly during tests recently conducted with a 278,000 DWT tanker to ascertain its manoeuvring characteristics at slow speeds in shallow water, it was found that the ship's master could reliably bring the ship to a stop from a manoevring speed of 3 or 4 knots with little or no change in the directional heading of the ship, and that this manoeuvring could be peformed while stopping the ship within a distance of a little over 3 ship lengths.

There is one other extremely important aspect to the speed and rudder control of vessels. When a ship is proceeding at a relatively high speed and has to take evasive or stopping action, inevitably the process of reversing the engines causes some loss of directional heading control because the forces on the rudder are influenced by the backing of the propeller. This is true whether the vessel be of single or twin screw. Accordingly the knowledgeable shiphandler, when faced with narrow or heavily trafficked waters, will generally select a rather slow manoeuvring speed which allows him not only to stop the ship in a very short distance while maintaining directional control, but also gives him the opportunity to increase his propeller speed which instantaneously increases his rudder effectiveness and thereby allows rather sharper turns than those described above without noticeably increasing the speed of the vessel.

Without dwelling for too long on the similarities or differences between ships and other transport vehicles, most of what has just been described about the critical importance of choosing a speed appropriate to the environment being encountered, is readily obvious to any driver or pilot faced with varying conditions.

Having described the basic elements of manoeuvring, two examples of successful practical manoeuvring situations with medium size and large tankers may be of some interest. Despite the fact that it is often suggested publicly that tankers cannot be predictably manoeuvred in either shallow water or at low speeds, many thousands of transits of canals, such as the famous Suez Canal, by both loaded and ballasted tankers have been accomplished for many years virtually without any serious incident or accident. This, despite the fact that the overall width of the Suez Canal is a scant four times the beam of the largest ship permitted to transit the canal, and the fact that the clearance between the bottom of the ship's keel and the bottom of the canal will at times be no more than 2 or 3 feet on a ship with a draught approaching 40 feet.

Another example of the excellent manoeuvrability of large and medium size tankers is the lightening operation (referred to later in chapter 6) which is routinely conducted in many parts of the world following its adoption first off the English south coast about ten years ago. In this operation a VLCC or ULCC will proceed at a speed as low as 2-3 knots while a smaller lightening ship in the range of 30-120,000 DWT moors alongside the larger ship without the assistance of tugboats or other aids. Obviously if the manoeuvring characteristics of these tankers were not entirely predictable and reliable, such operations could not be conducted with the excellent safety record that they have achieved while in the process of transferring many hundreds of millions of tons of oil.

One specialised manoeuvring device often fitted to smaller tankers deserves mention. This is the transverse or lateral thrusters most commonly fitted at the bow but in some instances at the stern as well on ships making frequent port calls and having to dock themselves. Bow and stern thrusters have proven their worth many times over in applications where the ships frequently need lateral forces to put them alongside, or to come off, piers unassisted. For the larger tankers with longer voyages and less frequent port calls, and with much greater levels of desired thrust not achievable with lateral thrusters, the installations have in the few test cases known proven to be of little value. Often it is suggested that bow and/or stern thrusters could provide emergency steering assistance for tankers or could improve the manoeuvring characteristics of tankers while routinely underway. Intensive model investigation and full scale tests by the author's company with bow and stern thrusters in ships of both the handy and VLCC sizes have repeatedly shown that at manoeuvring speeds much above 1 or 2 knots the effectiveness of bow and stern thrusters is reduced to virtually zero. Accordingly while these manoeuvring assistance devices have a valuable role to play for certain handy size tankers and for ships of other types, such as ferryboats and in dry cargo vessels, they should not be regarded as a panacea to be placed aboard all tankers regardless of size or trade.

Before leaving the subject of ship manoeuvring capability, one other matter should be addressed. Frequently it is suggested that 'brakes' should be put on tankers so as to make them able to quite literally stop on the spot. Investigations of manoeuvring or stopping assistance devices were conducted by Exxon a number of years ago to determine whether or not such equipment had any real technical merit. The results of these investigations are shown below for a twenty per cent reduction in stopping distance from 16 knots for a 200,000 DWT tanker.

Without having investigated the economics or structural requirements for any of these schemes, they can be characterised as expensive, not very effective and, in some cases even dangerous. Some represent essentially 'one shot' measures as well. Higher backing power is often suggested as a means for decreasing stopping distance. To be consistent with the table 7 and to achieve twenty per cent reduction in stopping distance from a speed of 16 knots would require an eighty per cent increase in a VLCCs backing power. While this is technically possible the same twenty per cent reduction in stopping distance could be achieved by simply reducing the initial speed from 16 to 14 knots!

While the data in table 7 relates to stopping from full speed, a more important measure of value involves stopping from a slow manoeuvring speed such as 6 knots. Approximate reductions in stopping distance for various design changes for a VLCC operating at 6 knots are shown in table 8.

Table 7: *Types of equipment to assist in stopping of tankers.*

Device	Specification
Water parachutes	Twelve 10-foot-diameter parachutes towed alongside the ship.
Water brake flaps	One flap on each side of the ship, extending perpendicularly out from the ship's sides like ears. Each flap 10 feet wide and 30 feet high.
Braking rockets	Total rocket motor thrust of 80,000 pounds for eight minutes. Could be accomplished with eight rockets operating consecutively for one minute each.
Braking jet engines	Four Boeing 707-type engines operated at near take-off condition. (These would have a total sustained thrust of 35 long tons or 20,000 pounds each).

Table 8: *Approximate reductions in stopping distance for various design changes for a VLCC operating at 6 knots.*

Propulsion Arrangement	Percentage reduction of stopping distance
Steam Turbine	Base Case
Double Astern Power	20–25%
Double Astern Power with Increased Propeller Diameter	Roughly 30%
Slow-Speed Diesel	About 20%
Kort Nozzle	15–20%
Controllable-Pitch Propeller	4%–6%

Table 8 indicates that there are a number of devices or concepts by which a noticeable decrease in stopping distance can be achieved. What it does not show however, is that a reduction in initial speed of less than ten per cent can achieve a more substantial decrease in stopping distance than can any of the above devices. Stated in different terms, if the ship entering harbour anticipates the need to stop promptly and reduces her speed from 6 to 5.5 knots the eventual stopping distance will be reduced by a greater value than any of the devices shown above. Once again it emphasises the critical nature of speed selection, not surprisingly a fundamental parameter for vehicles of all types.

What has just been described applies directly to manoeuvring both of tankers and ships of other types. It has been concerned with the technical characteristics and ability of the vessels themselves rather than the systems or rules under which ships must proceed while underway. In the author's view, however, it is these latter factors which are mainly responsible for safety of shipping while underway, particularly as regards vessel collisions and strandings, and the attendant fires and explosions which often follow collisions.

TRAFFIC CONTROL AND NAVIGATION

The fundamental rules under which vessels must proceed at sea are often called 'the rules of the road'. Such rules have existed for ships for many years, the most recent updating being the IMCO Collision Regulations of 1972 which came into force in mid-1976. These regulations describe the fundamental ways in which vessels meeting or passing each other must behave, and they describe as well the lights, sound signals and speed control which must be employed by vessels while underway under conditions of varying visibility.

An extremely important new concept was added to the basic 'rules of the road' system in the early days of IMCO. This is known as the Vessel Traffic Separation Schemes, over 100 of which are now in existence around the world. Figure 19 (pp110-111) shows the Vessel Traffic Separation Schemes applicable in the English Channel. It also indicates a deepwater route approved by IMCO to reserve deeper water needed for VLCCs and ULCCs approaching the Port of Rotterdam. In this regard, it is important to note that under the 1972 collision regulations, not only must all vessels in the vicinity of Traffic Separation Schemes abide by the Schemes themselves, but deepwater routes and the difficulty of manoeuvring of large vessels are both recognised as important elements in overall ship safety. In earlier days, no particular distinction had been made between large and small vessels or between deep draught vessels and shallower vessels more able to find manoeuvring space. These would not be particularly realistic concepts in today's world.

A variation of the basic Traffic Separation Scheme is so-called Advisory Vessel Traffic Systems or VTS which have increasingly been installed in major ports around the world. Shortly following the Second World War, particularly in England and on the continent, the value of Advisory Vessel Traffic Systems was recognised and implemented. Thus, with the enormous post-war growth of traffic entering and leaving the major ports of Europe, it was foreseen that the use of radar from central locations ashore, together with reliable radio communication with vessels proceeding in and out of harbours, could be used to advise vessels of impending dangerous situations. Furthermore, the shore advisory services in the hands of harbour authorities could be used to control the flow of traffic entering and leaving ports so as to minimise congestion and chaos which otherwise might result. The wisdom of these early decisions was proven many times over when, despite tremendous increases in numbers and sizes of the ships calling at major European ports, improvements in collision and grounding statistics were shown, almost universally, especially during periods of reduced or limited visibility.

On the ships themselves an equally substantial advance in navigational and collision avoidance ability has been produced by the nearly universal fitting of radar to ships. This particular development, however, was characterised by major growing pains known characteristically as the 'radar assisted collision'. There is no doubt whatsoever that in the immediate post-war years the supposition that radar would allow one to 'see in the dark' or in the fog, was manifestly incorrect. Statistics during the first ten years or so with radar commonly fitted to merchant shipping showed in fact that, relative to numbers of ships so equipped, many more accidents were occurring between ships in which both vessels had radar than in encounters between vessels where only one or neither of the vessels had radar. Nonetheless, once mariners had begun to appreciate not only the advantages but also the limitations of radar, its true value began to be understood. Properly applied and with the improved reliability which is slowly

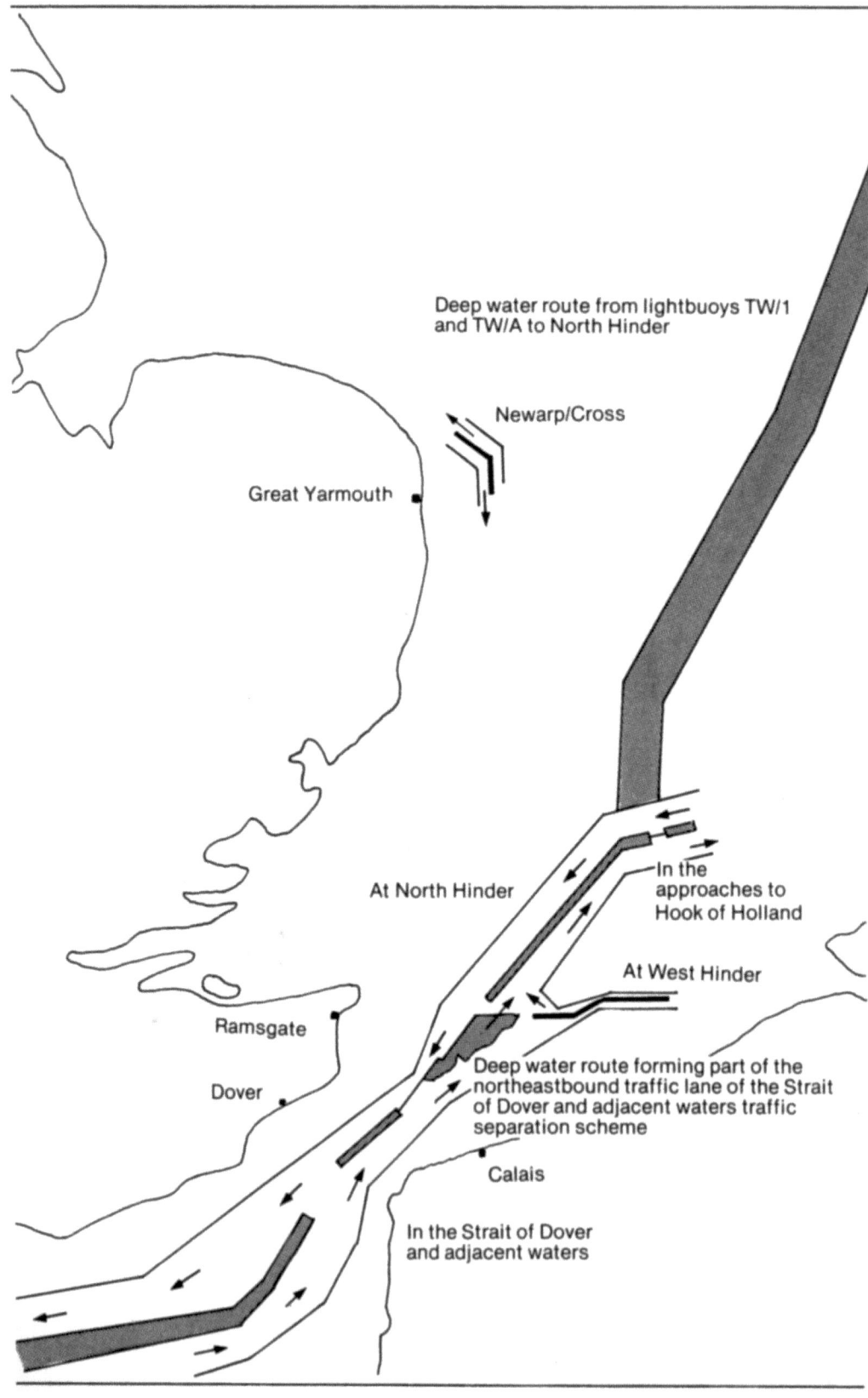

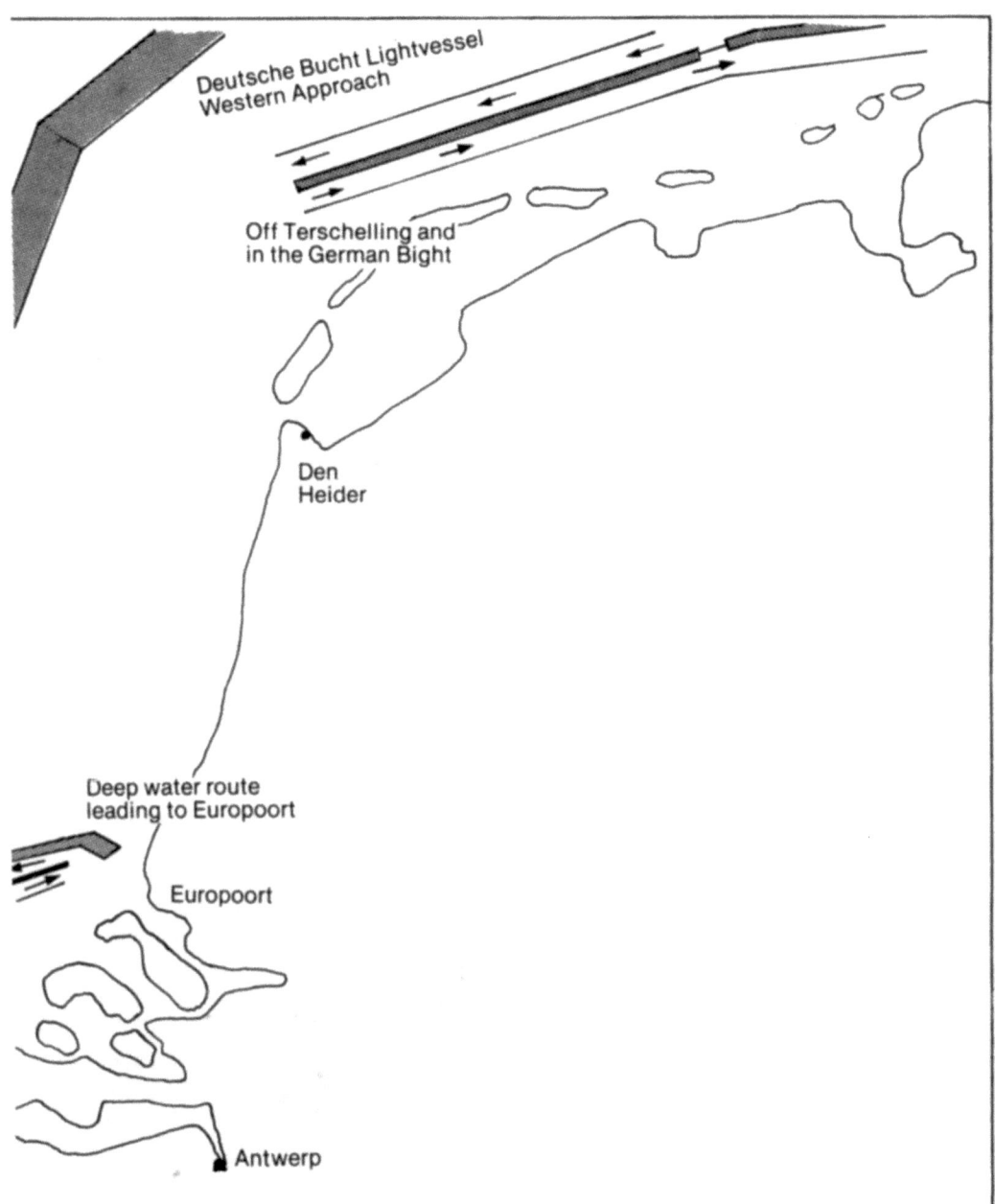

Figure 19: *Vessel Traffic Separation Schemes applicable in the English Channel.*

but surely being seen in marine radars, there is no doubt that radar can assist the collision avoidance activities of ships as well as assist in the basic navigation or position fixing of vessels.

A recent addition to the basic radar capability is so-called collision avoidance aids which free the bridge officers from manual plotting activities. While the author has no doubt that so-called collision avoidance systems or aids will became a substantial additional benefit to the ship's officers, they should be kept in proper perspective. Collision avoidance systems can do no more nor less than can the man himself in terms of plotting relative motion of other ship targets and indicating which targets may pose the most significant collision threats. Whereas a man with no other duties to occupy himself may be capable of concurrently plotting as many as three or four targets, the collision avoidance feature will track automatically upwards of a dozen or more targets, so the bridge officer remains free to conduct other duties and devote his time to assessing or judging the relative threat posed by various targets based on information from the collision avoidance equipment.

One other type of equipment which is of direct assistance to navigating officers not only on large tankers, but on shipping of all types, is the range of electronic communications and position-finding equipment which has become available in the last fifteen to twenty years. Three basic hyperbolic radio navigation position-finding systems exist in various portions of the world. These are known as Decca, commonly seen in UK, Northern Europe and the Middle East; Loran, which has been adopted for North America and Japan; and finally a system known as Omega, which is nearly worldwide at the present time but has somewhat lower accuracy than the other two systems. With any of these three systems it is possible for the navigator to determine his position with relatively good accuracy at any time of the day or night simply by obtaining readings from shore transmitters by a receiver aboard his ship, and adopting these to special charts showing line of position information for whichever system he is using. In reality, it is a pity that three systems, all of generically similar type, but none compatible in their actual reception equipment have been adopted as the fully equipped ship now must have at least two of these three systems if it is to be its prime navigation means around the world.

Perhaps the eventual navigation system of greatest utility in this respect will be that based upon tracking polar orbiting satellites launched by the US military establishment. So-called satellite navigation systems, despite growing pains as with other electronic systems, have been found increasingly useful in recent years. Despite these advances, however, and perhaps in no small measure because the reliability of electronic equipment on shipping has yet to reach the stage of development its manufacturers claim, navigators on virtually all well-found ships continue to rely upon celestial navigation, that is 'shooting' the sun and stars as their primary navigation system.

ACCIDENTS FROM NON-TANKERS

Most of what has been said already in this chapter relates to oil tankers and the roughly 200,000 tons of oil which it is estimated is spilled from tankers annually on a worldwide basis each year. It is worth noting however that by some estimates non-tanker accidents are believed to be responsible for an additional 100,000 tons of spillage each year. A brief indication of the factors contributing to this figure will round out the story of accidental oil spillage from vessels.

First, it must be appreciated that there are now believed to be over 50,000 merchant ships each over 100 grt in the world's fleet today. If one also includes smaller vessels beneath this figure including tugs, fishing craft and yachts, it becomes clear that there is a substantial population of vessels moving about the world's harbours, coasts and oceans. Just as with traffic situations for motor cars or aircraft, the smallest vessel can potentially put larger ones at hazard, especially in the more crowded areas in which less sea room is available for larger ships to manoeuvre. This of course is the fundamental reason why vessel traffic separation schemes and advisory vessel traffic systems are so vital in busy areas and have proven so effective.

Any motor-driven vessel using oil for fuel is a potential source of accidental pollution. For the majority of such vessels the size of potential spill is of course very small by comparison to that from a tanker. Nonetheless, the tremendous expansion in world trade has brought about major changes in the size and sophistication of merchant ships of all types. It is not surprising now to see 900 foot container ships with powers similar to that of large tankers and some container ships with over 100,000 horsepower and speeds in excess of 30 knots have been delivered putting them in a class with naval military vessels and major passenger lines. Ships having these general characteristics and burning several hundreds of tons of fuel a day carry an amount of bunkers equivalent to the full cargo of a small ocean-going tanker.

Thus, a collision or stranding of ships in this class also has a considerable potential for oil outflow. Fortunately as a result of the fact that most bunker fuel oil used in the marine world has had the volatile hydrocarbon components removed, these is little danger of fire or explosion as a result of rupture of bunker tanks as is the case with loaded crude oil or product tankers, but the pollution of the shore is just as bad.

CONCLUSION

This chapter has concentrated on the technical features of oil tankers inasmuch as these are so important when discussing oil pollution of the marine environment. Nonetheless, it must be stressed that the success of future endeavours to reduce oil pollution from shipping will be far more dependent upon the quality of the mariners on ships than on future changes in the ships themselves. It is manifestly impossible to prevent oil pollution by making ships strong enough to survive collision and grounding accidents without oil spillage, just as it is not possible to make airplanes crashproof. Accordingly with the adoption in July 1978 of a 'first ever' IMCO convention setting out international standards for training and certification of mariners we look for a continued improvement in the overall shipping safety record. This latest and perhaps most important of all IMCO conventions is a worthy companion to join previous IMCO initiatives on ships and traffic systems in a well rounded total system of international maritime safety regulations.

6

Operational pollution from tankers and other vessels

R. Maybourn
*Assistant General Manager
BP Tanker Co Ltd*

Because the major sources of crude oil are remote from the main areas of consumption, tankers, unlike dry cargo ships, to a great extent carry cargo in one direction only and, in order to navigate safely on the return voyage, carry substantial quantities of sea water as ballast. This applies equally to the product tankers, ships typically of up to about 30,000 tons cargo capacity designed for the carriage of petrol, kerosene, fuel oil and so forth and which distribute the various refined products from refineries to the customer. The need to maintain cargo tanks in a condition fit for the carriage of different types of cargo and also to carry water ballast in them on ballast voyages is at the root of what is known as operational pollution as distinct from pollution which may occur when oil is released to the sea as the result of a casualty. There is a much smaller, but nevertheless important, potential source of operational pollution which is unconnected with the cargo system and which applies to ships of all types. This is the engine-room and other machinery spaces where bilge water may become contaminated with fuel or lubricating oils. Finally another potential source of operational pollution of the sea is the accidental spillage of oil due to the over-flowing of tanks or the maloperation of sea valves.

* * *

Tankers vary in size from those which can carry a few hundred tons of oil to a handful which can load over half a million tons at a time. Certain features are common to them all. The hull is divided into a number of cargo compartments for reasons of stability, safety and convenience. Cargo is loaded into and removed from these tanks by a system of pipelines placed as near the ship's bottom as possible and which are connected to cargo pumps situated in a pump-room usually sited aft of, but immediately adjacent to the cargo tanks. On deck there are small steel hatches for each tank, vent pipes to control the passage of air or hydrocarbon vapours between the tanks and the atmosphere and some small access points such as those used for ullaging or sampling the cargo and so forth. (Oil quantities are commonly determined by measuring the distance from the top of the tank to the surface of the liquid. This distance is known as the ullage and the measuring procedure is called ullaging.) The cargo pumps are also connected to a discharge piping system on deck and to valves in the ship's hull which allow ballast water to be taken on or discharged. These various necessary features of the cargo handling system could, if operated incorrectly, allow oil to escape and cause pollution. The cargo compartments are not, however, the only spaces on board which carry oil. Substantial quantities of fuel oil are also carried, many thousands of tons in large ships, to cater for the needs of the boilers, main engines or other machinery on board. The bunker tanks which contain this oil are commonly placed both forward

Plate 12: *View from the bridge on board the BP tanker* British Respect.

and aft of the cargo tanks and as fuel is consumed during the voyage it is necessary to transfer oil from the forward tanks to those aft to supply the engine-room which is invariably placed at the stern of the ship.

When a tanker is at sea the main operations which could give rise to pollution if not conducted properly are the following:

       Ballasting
       Tank cleaning
       De-ballasting
       Transferring bunkers
       Discharging bilge water from machinery spaces

After its cargo has been discharged a tanker floats very high in the sea as the weights of the hull and machinery and the fuel oil on board are insufficient to immerse it very far. Typically the draught at the bow would be only a few feet and the propeller might be half exposed. In order to sink the hull to about half its loaded draught — a condition which permits efficient manoeuvrability — it is necessary to fill some of the cargo tanks with sea water to act as ballast. Since these tanks, and the pumps and pipelines handling the ballast water, contain oily residues from the cargo just discharged this ballast is contaminated and must not be returned to the sea until it has been suitably treated. At some stage on the ballast voyage a change of ballast has to take place as the water which is to be discharged at the loading port, prior to taking on a fresh cargo, has to meet a very high standard of cleanliness. This involves a period of intense activity, which can last many days on a large ship, during which tanks are cleaned to receive clean ballast, the pumps and pipelines are thoroughly washed and dirty ballast is discharged, all in such a way that the oily residues are retained on board and only a very small amount of oil in a very diluted form is released to the sea with the water being discharged.

The standards of pollution control which must be observed are defined in legislation. Efforts to control sea pollution date back to the 1920s, but even after World War II it was customary to discharge tank washings and other oily residues to the sea. The 1954 Oil Pollution Convention was a significant advance in that it established large sea areas adjacent to coastlines in which the discharge of oil was prohibited.[1] However, with the growth in the crude oil trade which has expanded nearly ten times since 1954, that convention has been substantially amended and strengthened. The amendments adopted by IMCO[2] in 1969 are those currently in force and these prohibit the discharge of oily

mixtures with any oil content at all from a tanker except under the following circumstances all of which must be satisfied:
- a the tanker is proceeding en route;
- b the instantaneous rate of discharge of oil content of any mixture must not exceed 60 litres per mile;
- c the total quantity of oil discharged on a ballast voyage must not exceed 1/15,000 of the total cargo carrying capacity;
- d the tanker must be more than 50 miles from the nearest land whenever an oily mixture is being discharged.

Not only have these amendments eliminated to a substantial degree the amount of oil which may be discharged as the result of essential operations carried out on the voyage but they ensure that the little oil which is released enters the sea in a form which permits it to be rapidly dispersed and degraded. The operational criteria just mentioned can only be achieved, however, if proper operational practices are followed. With this in mind the international tanker industry published two guides in 1973. These are the *Clean Seas Guide* and *Monitoring Load-on-Top*.[3,4] The first is for the use of tanker masters and crews and the second is for personnel at tanker loading terminals to enable them to check for themselves whether tankers arriving have complied with the Convention requirements whilst on the ballast passage.

TANK CLEANING AND LOAD-ON-TOP

It has already been mentioned that residues remain in cargo tanks and pumping systems after the cargo has been discharged. The quantities remaining in cargo tanks are very small after a cargo of refined oil but significant quantities (in terms of potential pollution) remain in cargo pumps and pipelines which can never be completely drained of oil. The residues remaining after crude oil has been discharged are generally very substantial, particularly in the cargo tanks, and since the volume of crude oil carried at sea is very large indeed, the need to control pollution as the result of operational requirements in this class of tanker is particularly great.

This need arises from the composition of crude oil which is a mixture of hydrocarbon compounds ranging from those which are very light and volatile to those at the opposite end of the spectrum. Some of the relatively heavy waxy and asphaltic components tend to settle out on the loaded passage, largely as a result of falling temperature which causes them to crystallise. They settle on the various horizontal members of the tank structure (which can be numerous and extensive in area) and they also form a layer on the tank bottom. These layers can be many centimetres thick on occasion and commonly amount to a quantity in the range of 0.3 to 0.5 per cent of the oil from which it has deposited. They remain clinging to the tank structure after the cargo has been discharged and, although they are refinable hydrocarbon material, they have traditionally been regarded as waste but in recent years they have been largely recovered as a result of the load-on-top procedure. In the typical large crude oil tanker they can amount to 1,000 tons or more with a further quantity of crude oil, perhaps 300 tons or so, remaining in the pumps and pipelines.

The ballast water which is necessary to enable a tanker to go to sea after discharging its cargo is pumped into selected cargo tanks. The residues remaining in those tanks contaminate it, however, as does the residual oil in the pumps and pipelines, hence its description, 'dirty ballast.' The ballast water on board when the tanker arrives at its loading port, and which will be discharged into the harbour, must be so clean that when it is pumped out no sheen or rainbow effect will be seen on the surface of the sea. This is known as the 'clean ballast' and there is no procedure available which will render dirty ballast fit to be discharged in this way.

Tank cleaning

Some cleaning of cargo tanks is carried out on every voyage for one or more of the following reasons:
    a   to prepare them for carrying clean ballast;
    b   to remove accumulations of oily residues;
    c   to render them fit for entry by personnel for inspection and maintenance.

The method most commonly employed is to use small portable machines having rotating nozzles which can be lowered into the compartment to be cleaned at the end of a hose supplying it with water at high pressure. In the very large crude oil tankers it has become common for the tank washing machines to be built permanently into the cargo tanks both to reduce labour and to permit much larger machines to be employed. Washing is carried out for as long as necessary and as it is being progressed a cargo pump is used to remove the wash water and the oil carried with it. This process is known as stripping and the washings removed from the tank are transferred to a slop tank system for processing.

Typically a pair of cargo tanks are used as the primary and secondary slop tanks in which gravity separation of oil and water takes place. Before tank cleaning begins both are flooded with sea water and are left connected to each other by an open pipeline. The secondary slop tank is also left connected to the sea through an open sea valve. As washing progresses the water and oil stripped from the cargo tanks are discharged into the primary slop tank at a point below but near the water surface. The oil floats to the surface and accumulates. The washing water, which is contaminated with fine particles of oil, disperses into the water already in the slop tank. At the same time water is displaced into the secondary slop tank and also to the sea. Both tanks act as separators and because of the long residence time before the washing water passes to the sea its oil content at this time will be low, perhaps of the order of 30 ppm.

When the necessary number of tanks have been cleaned, an operation that can go on for a number of days, a process known as line washing takes place. This is a procedure whereby sea water is sucked in by a cargo pump and discharged into the pipeline system to flush out any oily residues which may remain. Each pump is used in turn and valves are manipulated to ensure that every length of pipeline is cleaned. This takes some hours and uses a great deal of water.

This water is discharged into the primary slop tank and processed in the same way as washing water. When completed both pumps and tanks are clean and free from oil and ballasting with clean sea water can begin. The system is sufficiently clean at this time that the clean ballast taken in for discharge at the loading port will have an oil content of only about 5 ppm and visually will show no trace of oil whatsoever.

One final operation remains and this is to remove the dirty ballast taken in after the cargo was discharged. Since this ballast will have been on board for a number of days its oil content will, to a large extent, have gravitated to the surface which will be covered by a layer of oil. The ballast water itself will have an oil content of the order of 30 ppm and as it is discharged the pumping rate is controlled so that the rate of loss of oil to the sea does not exceed the permitted maximum of 60 litres per mile. This presents no problem for the bulk of the water in the tank. However, when the tank is nearly empty and only a few feet of water and its oily surface remain discharge overboard is stopped. These residues are transferred to the slop tanks and further separation of the oil from the ballast water takes place.

Load-on-Top

Although by this time the tank cleaning and ballasting operations have been completed the oil pollution control procedures have not. Both the slop tanks still contain contaminated water and the primary slop tank in particular contains also the large volume of oil recovered from the cargo system. Both tanks are allowed to settle for as long as convenient to reduce the oil content of the water before it is discharged to the sea. By this time it will contain typically around 100/150 ppm of oil. Discharge is carefully monitored to meet the discharge criteria and is stopped as the water/oil interface is approached. Any oil in the secondary slop tank is transferred to the primary which now has a water bottom above which is consolidated all the residues removed from the cargo tanks and pipelines. *(See figure 20.)* These residues are not pure oil, however, and contain a significant percentage of water in emulsion.

When the next cargo is loaded the slop tank can be isolated but more often it is not and crude oil is allowed to mix with the residues from the previous cargo. Hence the name load-on-top. This mixing often breaks the emulsion in the LOT material and releases much of the water which gravitates to the tank bottom. This may be decanted using a small pump and taking care to stop before the oil interface is reached or, alternatively, it may be discharged ashore into a slop tank before the bulk of the cargo is discharged. The LOT procedure has proved to be a most effective means of reducing operational pollution and of conserving oil which was formerly washed into the sea as waste. It involves little modification to the tanker since only the normal ship's tanks and piping are employed. It does, however, demand careful operation to obtain the best results and for this reason considerable research has been undertaken, particularly in the crude oil trade, to refine the procedure. The most important development is probably crude oil washing which has just obtained IMCO approval.

The objectives of load-on-top as a pollution control measure can be put into perspective by quoting from the booklet *Monitoring of Load-on-Top*.

> It is important to recognise that the permitted discharge is only about 0.006% of the cargo dead-weight, equivalent to about 1% to 2% of the oil that remains in the tanker after completion of a full cargo discharge. This means that 98% to 99% of the oil that remains on board at the end of discharge must still be retained on board after completion of tank cleaning procedures and change of ballast. This can be accomplished by using the Load-on-Top procedures that are detailed in the *Clean Seas Guide for Oil Tankers*.

122 . Prevention of Oil Pollution

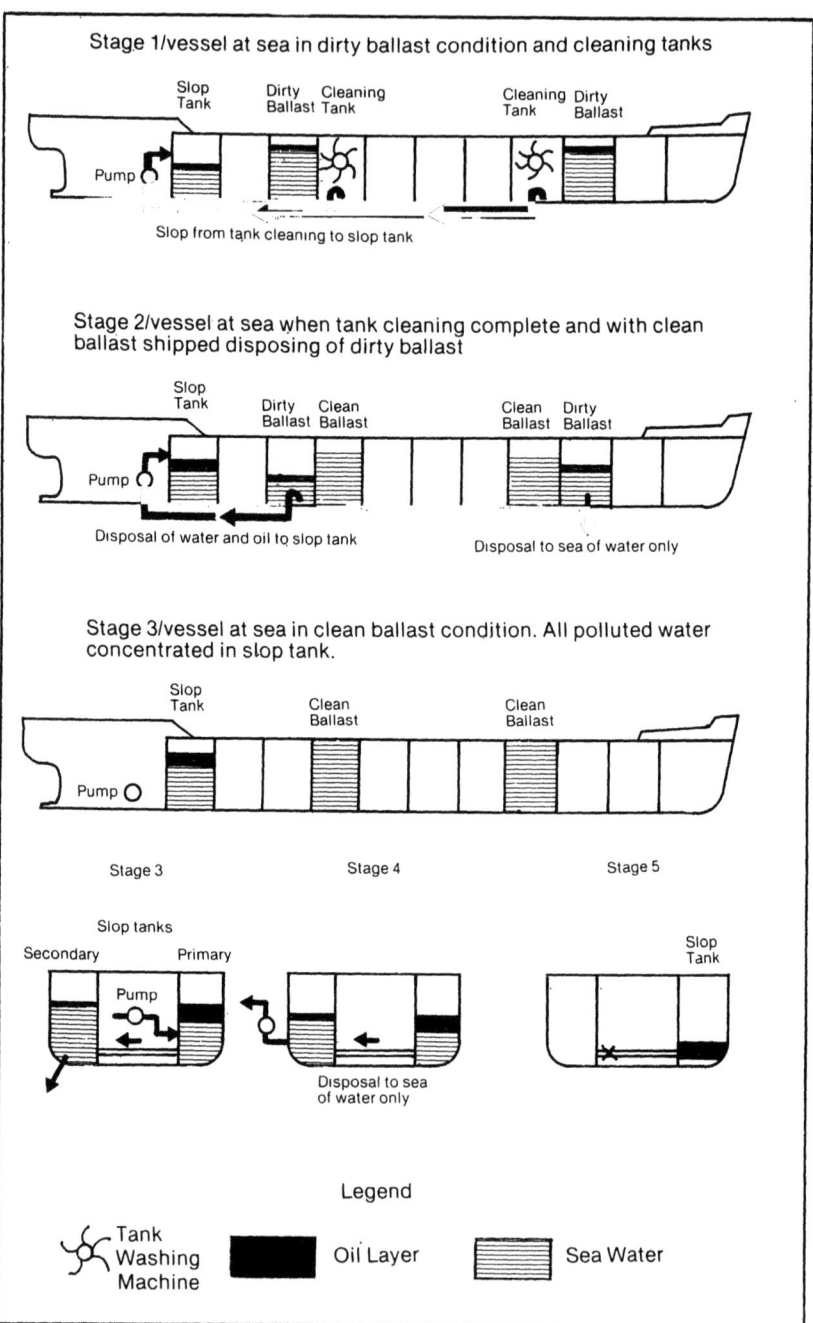

Figure 20: *32,000 DWT crude oil carrier using load-on-top system for anti-pollution*

Crude oil washing

The condition of the cargo tanks after a crude oil cargo has been discharged has already been described. It was recognised many years ago that water is not an ideal medium for washing tanks and that the contamination of the water employed causes pollution control problems. It was also recognised that crude oil itself might be better but no safe or practical means of employing it existed.

The advent of the very large crude carrier (VLCC) with cargo tanks having capacities in some cases in excess of 25,000 tons each demanded a re-appraisal of tank cleaning methods. The large capacity fixed-in-tank washing machine was developed which could be supplied with its washing fluid by the cargo pumps. It thus became possible to use crude oil as a washing medium and experimental work to this end began about 1972. The system as developed is employed only in ships having an inert gas system which ensures that a flammable atmosphere cannot exist in a cargo tank. This is necessary because large capacity washing machines can, under certain circumstances, create charges of static electricity capable of incendive discharge.

Oil discharged from a tank is usually replaced by air which mixes with any vapours already present. If the oil being discharged is volatile the air/vapour mixtures which are formed may be flammable. A ship equipped with an inert gas system uses flue gas from the funnel uptake to replace the cargo being discharged. This flue gas has a very low oxygen content (less than five per cent) and will not support combustion. It is known as 'inert gas' and inert gas systems are widely fitted in VLCCs. They will become compulsory in due course for all crude oil tankers down to 20,000 tons cargo capacity.

The object of crude oil washing is to redisperse the residues which fall out during the voyage back into the cargo. This is achieved by operating the tank cleaning machines to a predetermined pattern using crude oil diverted from the cargo being discharged ashore. A typical washing pattern involves washing the upper part of a tank when half discharged and then resuming at the latter stage of discharge to clean the lower tank structure and the tank bottom. Crude oil has proved to be a much more effective medium than water and the cleanliness achieved is such that the tank resembles one which has previously held a refined oil such as gas oil or marine diesel.

The washing process must, of course, be carried out during cargo discharge and this has the important benefit that the residues are transferred to the shore tanks where they properly belong. The crude oil tanker having crude oil washed is thus in a condition not dissimilar from that of a products tanker having discharged a refined oil. By paying careful attention to draining cargo lines and pumps the oil remaining in the system can be reduced to 100 tons or so for a VLCC compared with the 1,000 to 2,000 tons which may remain if crude oil washing is not practised.

Before sailing it is necessary to take on dirty ballast in the normal way but as the tanks into which it is placed are free from residues the ballast water becomes only lightly contaminated with some of the oil which could not be stripped from the system. After sailing the clean ballast tanks are prepared for ballast. They require only a rinse with water to remove any loose oil which may remain near the strum and to flush any oil in the suction piping back into the slop tanks. This is a quick and easy operation after which the lines and pumps are washed prior to taking on clean ballast. No other tank cleaning is required. After the clean ballast has been pumped in the dirty ballast must be removed. Its oil content is low and the residual oil on its surface is commonly only a few centimetres thick. Nonetheless the LOT procedure is carried out as with water washing to ensure that none escapes into the sea. The oil retained in the slop tank is about only 10 per cent of what would have been the case had crude oil washing not been carried out thus reducing the potential for pollution and making the pollution control procedures easier to carry out.

The benefits of the system are such that it was accepted by IMCO in February 1978 as an alternative to segregated ballast tanks. Segregated ballast is a concept which requires ballast water to be carried in tanks dedicated for this purpose and which prohibits, except in exceptional circumstances, the carriage of ballast water in cargo tanks. To assist tanker personnel in understanding the technique the industry published a booklet in 1977 entitled *Guidelines for Tankwashing with Crude Oil*.[5]

Engine room bilge water

The bilges of engine rooms and other machinery spaces such as pump rooms invariably contain some water and oil. This water may enter, for example, through the propeller shaft stern gland or accumulate as leakage from pumps or other equipment. It may also result from cleaning operations when hoses may be used to wash down oil fuel leakage from the boiler room or lubricating oil leakages from the main engine or other machinery.

This contaminated bilge water is an embarrassment in port as it must not be discharged into the sea where it is likely to cause visible pollution. Good housekeeping demands that every effort is made to minimise leaks of both oil and water so as to keep the problem under control. Where a stay in port is prolonged, however, the accumulation of bilge water may be such that some action is essential and tankers are often provided with a discharge line from the bilge pump to a cofferdam where the contaminated water may be held for subsequent processing at sea.

The 1973 Pollution Convention contains strict provisions regarding the discharge of bilge water. Although not yet ratified some governments have already passed domestic legislation in line with the convention and ICS has encouraged owners to implement such provisions as are practicable immediately on a voluntary basis. In such cases bilge water discharge is permitted only when a vessel is proceeding en route and is more than 12 miles from land and if it has been treated or separated so that its oil content does not exceed 100 ppm.

OIL TRANSFER OPERATIONS

The operations described so far all concern the contamination of water with oil within a ship as a result of various day to day activities. There are, however, other operations which may be carried out at sea involving the movement of oil only and which have the potential for serious pollution if not conducted properly.

Bunker transfer

The fuel burned under the boilers or in the engines is taken from settling tanks in the engine room where it is treated to remove water and unwanted sediments. These tanks are topped up at regular intervals from the bunker tanks which are commonly sited between the cargo pumproom and the engine room. On long voyages it is often necessary to transfer oil from the forward bunker tanks to those aft to replace the oil consumed and sometimes also to prevent the ship trimming too far by the head. This transfer operation is carried out using the bunker transfer pump which is placed in the small pumproom adjacent to the forward bunker tanks. A bunker line connects the various bunker tanks and it also has branches with valve connections at the cargo loading and discharging manifolds. The most common cause of spillage during transfer is overfilling the tanks receiving the bunkers, often because the operation takes some hours and the pumping rate has been misjudged. It is not uncommon nowadays for high level alarms to be fitted to bunker tanks and these give an audible warning when the oil reaches a pre-selected level which allows ample time for the operation to be stopped safely. Other potential sources of spillage are failure to shut the bunker manifold valves or leakage from the line because of corrosion or through the expansion joints. The latter, however, would be very slight.

Cargo transfer

Once loaded it is normal practice never to move cargo from tank to tank and the loading ullages and quantities are entered on the bill of lading. Exceptionally, however, cargo may be transferred at sea. This may arise in the unlikely event of a leak developing or, in some large crude tankers, because it is necessary to do so to maintain the ship on an even keel to minimise draught at the arrival port. To transfer cargo the ships' internal pipelines and a cargo pump are used. The operation requires care to ensure that the correct valves are employed and close supervision to prevent an overflow as the tanks receiving the oil are likely already to contain cargo and have quite small ullages.

Ship to ship transfer

Oil is quite often transferred from ship to ship or ship to barge in harbours or estuaries. In recent years the practice has developed of transferring cargo in the open sea from VLCCs to smaller tankers mainly to allow them to enter ports for which their draught would otherwise be too great. This has become, in many cases, a sophisticated operation employing lightening tankers specially equipped with large pneumatic fenders and cranes for handling the cargo hoses. *(See plate 13)* The operation is carried out in selected areas where there is reasonable shelter from bad weather and ample room to manoeuvre. Because of the specialised nature of the transfer operation and the pollution which could arise if not carried out efficiently OCIMF and ICS have jointly published a Guide which contains recommendations on:
    a Safety standards;
    b Minimum equipment standards for ships for both offshore and sheltered water transfers.
    c Sound operating practice.[6]

Plate 13: *Lightening operation,* British Dragoon, *53,000 grt, alongside a VLCC.*

More often than not the transfer takes place in territorial waters and the guide serves as a useful basis for agreement with government authorities when agreeing the operating procedures to be followed.

There are three aspects of the operation which have, if not properly handled, the potential to cause pollution. These are the berthing of the two ships alongside one another where hull damage is the risk; the connection of hoses, ullaging etc during which safety precautions are observed to prevent the ignition of oil or vapours; and the transfer of oil itself.

Berthing often takes place with both ships underway and the large loaded ship maintains course and speed while the smaller tanker manoeuvres alongside using its fenders to absorb the berthing impact. After mooring both ships stop and come to anchor. Once moored together strict safety procedures have to be observed by both ships. Crew are standing by to disconnect hoses and unmoor at short notice, smoking regulations are enforced, transmitting aerials are earthed and so forth. The guide contains a safety appendix giving advice on such matters as the use of cathodic protections systems, insulating moorings, cargo hose bonding and, most importantly, the check lists. There are four of these as follows:

1. Before operations commence
2. Before run-in, berthing and mooring
3. Before cargo transfer
4. Before unmooring

and each has to be completed by both ships. They are an effective means of ensuring that the requirements of the operation are checked and fully understood.

The oil transfer is basically a normal discharging operation for the VLCC and a normal loading operation for the transfer tanker. However, if the transfer tanker has been modified and is regularly employed lightening ships she may well have segregated ballast tanks or retain cargo on board as ballast. If not she will either clean tanks alongside her discharge berth or retain the dirty ballast on board while loading her cargo. The discharge of ballast other than that which is clean is strictly prohibited.

HOW AND WHY DO SPILLAGES OCCUR?

There are two reasons why spillages which cause pollution occur; mechanical failure and maloperation. Mechanical failure embraces such incidents as hose failure, hull leakage and defective ships' side valves. Sometimes, though by no means always, it will be the result of poor maintenance or maloperation. Maloperation, however, more generally embraces a range of human failures arising from inattention, inexperience and, not infrequently, inexplicable failures by responsible and experienced personnel. It results in overflows, accidental discharge of oil to the sea, failure to observe properly the load-on-top procedure and so forth. The high incidence of leakage through sea valves causing pollution while handling cargo alongside loading and discharge berths caused OCIMF and ICS to carry out a detailed study into the problem. They subsequently issued a guide giving advice on how to avoid pollution from this source.[7] This contains much information about such valves and gives advice on testing, securing them during cargo discharge and opening them for ballasting.

The need to train tanker personnel to carry out their duties effectively is receiving increasing attention at IMCO and elsewhere and it is significant that the recent IMCO Tanker Safety and Pollution Prevention Conference set out strict requirements both for guidelines and officer experience before being allowed to operate crude oil washing.[8] To comply fully with the various provisions of anti-pollution legislation places an onerous burden on ships' personnel. To assist them it has become common in recent years for managements to provide comprehensive operating instructions supported in many cases by check lists listing the essential elements of the various procedures which must be followed. Senior officers are also encouraged to pre-plan cargo, ballast and tank cleaning operations and to provide detailed written instructions for their subordinates. Check lists and pre-planning help to avoid what is sometimes called 'competent failure' by experienced personnel and the existence of these aids is a useful indication to visiting personnel from the shore of an efficient and well run ship.[10]

Controls

Although the current legislation which governs pollution prevention is comprehensive and strict when a ship is at sea it is left to ships' officers to comply without external supervision. However, the International Convention for the Prevention of Pollution of the Sea by Oil requires tankers to carry an oil record book in a specified form. Comprehensive details of tank cleaning, ballast changing and so forth must be entered with dates, times and the ship's position and the record book is regularly inspected by the maritime administration of the flag state. Severe penalties can be applied if the book is not properly maintained or there is evidence that pollution has taken place.

In order to overcome this lack of direct control the 1973 Pollution Convention has written into it requirements for onboard control of the 'black box' type. The prime requirement is an oil in water monitor capable of measuring the oil content of any water being discharged to the sea. Associated with it is a flow meter measuring the effluent discharge rate and the measurement of ships' speed. These three inputs are continuously recorded and the 'black box' will check and control both the instantaneous rate of discharge and the total quantity of oil discharged to the sea. In the event of the maximum permissible oil loss of deadweight/15000 is reached the pumps and sea valves will be closed and no further discharges will be permitted. Difficulties in developing suitable oil in water monitors have delayed the ratification of the 1973 Convention which also contains other significant advances in oil pollution control. In order that the benefits of the measures which are capable of implementation should not be lost the ICS introduced in 1976 a code, commonly known as the 'Voluntary Code.'[9] Owners are invited to observe those parts of the 1973 Convention which can be implemented and which are detailed in the code, and to register their acceptance with ICS. This has been done by a large number of companies. One other practical and effective check on the pollution control procedures carried out on the ballast voyage is to measure the residues remaining on board prior to loading the next cargo. These should be appropriate to the amount of tank cleaning carried out and guidance for personnel at the loading port is contained in the OCIMF/ICS booklet *Monitoring Load-on-Top* mentioned earlier.

THE FUTURE

A significant provision of the 1973 Convention relates to new tankers of over 70,000 tons cargo capacity. These vessels are required to carry their sea-going ballast in special tanks dedicated to this duty known as segregated ballast tanks, and ballast can only be carried in cargo tanks in exceptional circumstances. This requirement reduces the cargo carrying capacity by about 15 per cent but it substantially eliminates operational pollution as it occurs in most existing tankers because tank washing for clean ballast purposes is not necessary. Additionally the loss of oil to the sea permitted in segregated ballast tankers as a result of tank cleaning operations is reduced to deadweight 30,000 compared with twice that amount allowed for conventional tankers. Very few ships have been built or ordered, however, which would have to comply to these requirements.

Following a series of tanker accidents in 1976 and 1977 and pressure from the US Government an IMCO Conference (Tanker Safety and Pollution Prevention) was held in February 1978. An outcome of this conference was the addition of a protocol to the 1973 Convention and a strong recommendation that both the convention and its protocol should enter into force by 1981. So far as operational pollution is concerned, the most significant change in the convention is the extension of the segregated ballast concept down to new ships of 20,000 deadweight tons if crude oil carriers or 30,000 tons otherwise. Existing tankers did not escape scrutiny either and such ships have the options of retro-fitting segregated ballast tanks, or adopting 'clean ballast' or converting so as to be able to carry out crude oil washing. 'Clean ballast' is a concept similar to segregated ballast but structural changes are not necessary. However, it is to be permitted only as an interim measure in crude oil carriers pending the adoption of either segregated ballast or crude oil washing. All crude oil tankers above 40,000 tons deadweight will have to adopt one of the three options from the date the convention enters into force. Within four years of that date the clean ballast option will be removed. Products tankers will have to adopt comparable measures also but in their case all vessels over 40,000 tons will have the option of adopting either the 'clean ballast' concept or converting to segregated ballast from the date of entry into force.

These changes are important since the traditional tank cleaning procedures will be greatly simplified well within the next decade substantially eliminating those parts of the pollution control process most difficult to control and requiring the greatest care. It is to be expected that owners will generally adopt crude oil washing in preference to retro-fitted segregated ballast since it is cheaper and does not reduce the cargo carrying (and earning) capacity of the ship. Control will be easier for governments since a tanker will only be certified for crude oil washing after it has demonstrated that its tanks can be cleaned of residues during cargo discharge and its cargo lines and pumps can be drained of oil.

Spot checks will be carried out at discharge ports to ensure that the required standards are being met.

REFERENCES
1 The relevant conventions are: International Convention for the Prevention of Pollution of the Sea by Oil 1954, amended in 1962 and 1969.
2 IMCO, properly the Inter-Governmental Maritime Consultative Organisation, is the UN agency responsible for maritime safety. It was established, with its headquarters at 101 Piccadilly, London W1, in 1959.
3 Oil Companies' International Marine Forum and International Chamber of Shipping, *Clean Seas Guide for Oil Tankers (The Operation of Load-on-Top)*, (Witherby & Co Ltd, Aylesbury).
4 Oil Companies' International Marine Forum and International Chamber of Shipping, *Monitoring of Load-on-Top*, (Witherby Co Ltd, Aylesbury).
5 Oil Companies' International Marine Forum and International Chamber of Shipping, *Guidelines for Tank Washing with Crude Oil* (Witherby & Co, Aylesbury).
6 Oil Companies' International Marine Forum and International Chamber of Shipping, *Ship to Ship Transfer Guide (Petroleum)* (Witherby & Co, Aylesbury).
7 Oil Companies' International Marine Forum and International Chamber of Shipping, *Prevention of Oil Spillages through Cargo Pumproom Sea Valves* (Witherby & Co, Aylesbury).
8 International Convention for the Prevention of Pollution from Ships, 1973 (also known as *MARPOL* 1973), protocol added 1978. (*MARPOL* 1973 is not yet ratified, but there is a target date of 1981 for it and its protocol.)
9 International Chamber of Shipping, *Pollution Prevention Code* (International Chamber of Shipping, London).
10 There is a very good, short IMCO publication setting out the development of anti-pollution legislation: Yoshio Sasamura, Director, Marine Environment Division, IMCO *Environmental Impact of the Transportation of Oil*.

7

Port operations: spills during loading, discharging and bunkering

R S Hawkins
Texaco Overseas Tankship Ltd

Once a tanker enters port risks of pollution proliferate. Not only is the potential for environmental damage all the greater but the consequences are more visible. It is a period of intense activity which puts a high work load on both personnel and machinery. In the case of a typical VLCC trading pattern, for instance, her turnaround might take only thirty-six hours after a sea passage lasting more than a month. Moreover, port operations are more exposed to the public gaze than incidents on a long ocean passage.

An essential feature of tanker port operations is the transfer of oil between the vessel and the shore. For this to be accomplished safely and without pollution the vessel has to be securely moored and a temporary and portable connection must be made between the two pipeline systems. (*See figure 21*) It is also likely that the cargo will not be the only oil flowing between the ship and the shore; bunkers must be taken and even lubricating oil is often loaded in bulk. At the same time many other activities will be in full swing: stores and spare gear will be delivered and must be checked and stowed away, crude oil washing of tanks might be in progress, repairs and routine maintenance will certainly be undertaken, surveys and ship inspections might be carried out and, on top of all this, a proportion of the ship's personnel is likely to be changing. It is a time of hazard and exposure yet also a period when the ship's structure and equipment is vulnerable and her crew under stress.

To an observer it might seem that more time and effort is spent on these servicing activities than on the actual transfer of cargo. In terms of manpower alone such an impression might well be true but, in terms of care and concentration, it is certainly not. Efficient cargo operations require diligence rather than bustling effort; they are attention demanding rather than labour intensive. The importance of this will be appreciated when it is remembered that something in the region of three-quarters of all pollution incidents result from human error.

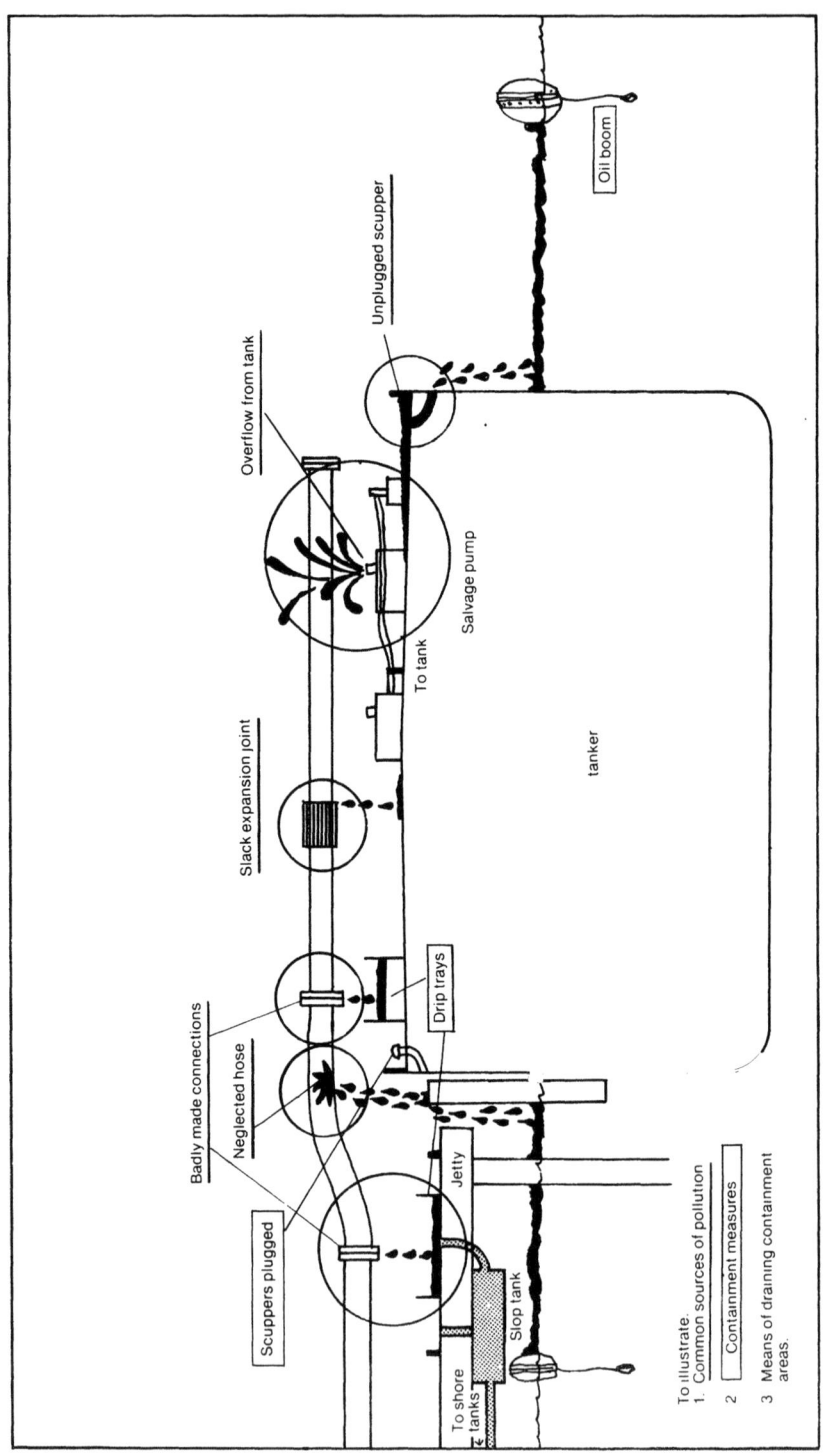

Figure 21: *Possible sources and remedies of oil pollution during port operations.*

In the light of all this, it need come as no surprise to find that tanker port operations are covered by a multitude of laws and regulations drawn up by governmental and international agencies and supported by the force of a considerable body of public opinion, the legal aspects of which are covered in a separate chapter.

CARGO OPERATIONS

For those readers who are not hardened tankermen it might be helpful at this stage to describe briefly the main cargo operations undertaken in port and to show how they might lead to pollution risks.

At each end of the voyage the operations are opposite in purpose, but not so contrary in nature. Cargo must be loaded before it can be transported and discharged, so consideration of the voyage can be started at the loading port.

Loading

The tanker will arrive with sufficient clean ballast for safe manouvring and, once berthed, this will either be discharged directly overboard or, if the local regulations require it, ashore. Though, in each case, cargo can be loaded simultaneously, provided that the pipeline and valve system allows adequate separation. The accepted standard is to have at least two closed valves between oil and ballast water. Once the hoses are connected and the valves are set, the flow of oil into the ship can commence — and the risk of pollution also commences!

From that moment on defective or malfunctioning equipment, a moment's lack of attention, or any other human error might allow that oil to escape from the confines of the tanks and pipelines. Tracing the flow between the jetty and the vessel's tanks will show where the greatest risks occur. The cargo hoses are suspended over the water, in the event of a leak it will enter the water. After crossing the rail they are connected to the vessel's manifold and, if the connection is not properly made, then oil will leak at the joint.

Once the oil reaches the tanks there is little risk of pollution until the critical stage of 'topping-up' is reached except in the event of structural failure. It is at this stage that the level reaches its maximum close under the deck and an overflow becomes possible through faulty equipment or, more likely, human error.

Discharging

At the other end of the voyage the oil must be pumped from the vessel's tanks, through the flexible hoses, across the jetty and into shore tanks. When considering pollution risks this operation is not so different from loading; the oil is flowing in the opposite direction but through similar pipelines. And, strange as it may seem, it is still possible for a cargo tank to overflow during discharge operations. Also it is as well to remember that if the vessel is bunkering, there are in fact oil tanks being filled and 'topped-off' and they can through a moment's inattention overflow just as easily as cargo tanks.

At some stage of the discharge operation ballast will have to be taken on board. In the case of larger vessels this is done while a proportion of the cargo remains on board and, even when it is done on completion of discharge, there is still likely to be oil remaining in the pipelines. So this is a time when the correct procedures must be carefully and strictly followed, for the vessel's integrity is deliberately broken by opening the sea-valves. Although the sea-chests are situated at the bottom of the pumproom, well below sea level even at the lightest draught, oil has a nasty habit of escaping in defiance of all laws of hydrostatics. The standard instruction has always been that the sea-valve should not be opened until the pump is running, but a pump might fail to pick up suction at the critical moment and if it does turbulence in the sea-chest could allow oil to escape against the inward water pressure. So in practice a more sophisticated technique is needed. If ballasting is to be started without any risk of pollution a vacuum must be obtained on the sea-suction before the sea valve is opened. The following recommended procedure does this and, at the same time, avoids having to start large main cargo pumps 'dry' (which is obviously unsatisfactory and has led to mechanical failures of the pumps).

After lining up the appropriate valves in the cargo system to take on ballast the first stage is to pump out the sea-suction lines using the stripping pump (smaller, often reciprocating pumps used to complete the discharging operation). Valves between the main pump suction and the outer sea-valve should be opened successively until a good vacuum is obtained. Suction can then be transfered to the main pump casing through the priming line and the suction valve opened. With the stripping pump still maintaining a vacuum the outermost sea-valve can now be opened slightly in order to fill the line and main cargo pump with water. The stripping pump itself should be discharged into the slop tank throughout these procedures. There are several advantages in this technique. The flow of water is not so great and, therefore, any turbulence is that much less violent, and the main cargo pump is full of water before it is started. When this is now done and the sea-valve is fully opened there will be little chance that it will fail to pick up suction. And once it is confirmed that ballast is being delivered to the tanks the stripping system can be shut off and the pump stopped.

On completion of discharge the manifold connection must be broken before the hoses are lifted back ashore. But first they must be thoroughly drained and the open ends closed with blank plates so as to prevent oil succumbing to gravity and draining back onto the deck or dripping directly into the water.

The ship is then ready to sail and to face the pollution risks of the passage, but these are the subject of another chapter.

Mooring

The transfer of oil between the ship and shore requires the tanker be safely moored. That all jetties and mooring facilities can do this in prevailing local weather conditions is the basic design criterion. Incidentally, it is necessary to refer to 'mooring facilities' as tanker port operations do not require a conventional jetty; at many ports buoy moorings with floating or submerged pipelines can be used, and in trans-shipping operations one ship becomes a 'jetty' for the other.

Beyond this basic requirement many other considerations will influence jetty design. Such factors as tidal range and strength of current will need to be considered in addition to the size of the ship it is intended to berth. Not only must the jetty be strong enough to hold the vessel alongside but also it must be able to withstand the impact as she comes alongside. In the case of the largest vessels their inertia is so great that the speed of approach must be carefully restricted in order to prevent the jetty being completely demolished.

While loading and discharging a tanker's freeboard will vary considerably, the tides will rise and fall, and she is bound to range along the jetty to some extent. For all these reasons there must be considerable flexibility in the connections between the ship's and shore pipeline systems. In the past this was always achieved by the use of reinforced rubber hoses which could easily become a pollution risk. Their condition can deteriorate without visible signs and negligent handling can easily and quickly damage them to failure point. In more recent years several patent forms of rigs have been developed which either ensure that the hoses are handled properly or avoid their disadvantages altogether. This latter type have steel arms provided with oil tight, universal jointed, swivel couplings and these are better able to withstand the higher pressure generated by the fast loading and discharging rates of modern ships.

A short description of a typical tanker berth should help to make the above description clearer. It is a concrete pile structure capable of berthing vessels of from 50,000 tons deadweight up to the largest vessels afloat today. Yet the jetty itself is no more than a third the length of such a vessel; when working tankers there is just no need for a large cargo handling area. The outlying points that are necessary to give the vessel's moorings sufficient spread are provided by isolated dolphins of considerable strength and similar in construction to the jetty itself but only connected to it by light walkways. There are three cargo handling arms each with a diameter of 16 inches, with a fourth, 12 inches in diameter for bunkering. This rig is capable of providing ample flexibility as the ends of the arms are able to reach from 4 to 30 metres above sea level and from 3 to 11 metres out from the face of the jetty. In addition they are capable of moving 4.5 metres either side of their centre line. These limits enable them to cope with vessels whose freeboard might vary from 3 to 18 metres.

Although the pile construction provides a strong yet flexible structure and a fendering system is provided along the face of the jetty and dolphins to cushion the impact, the speed of approach of the larger vessels that use it will need to be restricted. For instance a 100,000 ton deadweight vessel could safely make contact with a rate of approach of 8 metres a minute while a 250,000 tonner must reduce this to below 5 metres per minute. Such small and precise speeds cannot be judged by eye and on modern jetties some form of radar to measure the speed of approach is usually used.

Once alongside, the vessel is held by a system of lines whose general arrangement is as old as ships themselves. Nothing better than the age old web of headlines, sternlines, breastlines and springs has yet been found. Only, in the case of large tankers most of the lines will be of steel wire, over 6 inches in circumference, and wound permanently on winch drums for ease of handling. Their strength is matched by the quick release hooks on the dolphins each of 150 tons safe working load.

Plate 14: *Loading product at the Shell refinery, Singapore, into the Koratiya.*

Plate 15: *Remote controls on the tanker end of a chiksam (rigid arm, oilflow boom) unloading a VLCC.*
Plate 16: *Hydraulic flange coupling, which obviates the need to fit all the flange bolts.*

The 'tools for the job'

While there is a gallon of oil on board the ship there remains a risk of pollution. Being a liquid it must always be contained in drums, in tanks or in pipelines. While it is safely within these containers there can be no pollution — the problems arise when it escapes. The ship and her equipment are the only tools available to prevent pollution.

The most effective tool is the structure of the ship herself. Her hull contains thousands of tons of oil; up to 500,000 in the case of the very largest vessels. Fortunately such amounts are not held in one vast tank and recent legislation has sought to further limit the size of individual tanks so as to reduce the amount of pollution in the event of structural failure.

While oil is being transported there is little risk of pollution except in the event of the hull being damaged. However, we have seen that in the course of tanker operations it becomes necessary to transfer it between the ship and the shore. Once this is being done the opportunities for it to escape are multiplied. In fact the situation is made even worse as, in the course of pumping, considerable pressures can be generated in the pipelines. A small seep at a joint can quickly and suddenly become a flood — or a spectacular black fountain.

So it is important that the tools be kept in first class condition; valves and pipelines must be tight and it is a regular routine to prove this on ballast passages. The flexible rubber hoses, or the universal joints on the patent rigs that connect the ship and the jetty will be subjected to just the same pressure as the steel pipelines and so must not be a weak link. They must be regularly inspected and pressure tested. When in use hoses should be properly suspended in cradles, or at least with slings that will not cut into them, so as to avoid sharp bends or kinking. The supports should be adjusted so as to prevent their chaffing on any sharp edges at the ship's side and to avoid excess weight being put on the manifold. They will need attention and adjustment throughout the transfer operation as the relative level of the ship's deck and jetty change. Many of the hydraulically controlled rigs can sense the changing conditions for themselves and make the necessary adjustments automatically.

The vessel's manifold has already been pinpointed as one danger point. The flanges must be clean and flat, the packing properly cut and inserted, and all the bolts used and tightened. It is equally as important that the blanks on the opposite side manifold have had similar attention and be just as carefully fitted. So often they are forgotten, but they can so easily become the last line of defence against leaking valves.

Many of the commonest oil leaks during port operations, leaking joints at the manifold and overflowing tanks for instance, tend to spill oil on to the deck. Dangerous and messy as that is, it need not necessarily result in pollution. If the principle of containment can be applied the oil will not escape into the environment; in fact the ship is constructed and can be prepared so as to provide the necessary containment. However at sea just the reverse is necessary; scupper drains must be provided to clear quickly the water that is shipped. Therefore, the first anti-pollution precaution before any oil transfer operations commence is to make sure these are plugged in order that the raised gunwale bar can prevent oil from running straight over the side and into the water. In the event the situation is frequently less straightforward than that. Heavy rain might fill the deck to overflowing even before any oil is spilled, or the flow might be sufficient to do the same in itself. So it is usual to provide a further precaution by having a portable pump ready to transfer accummulating oil back into a slop tank.

In the manifold area, containment of small spills can be even more local. Either fixed or portable drip-trays are placed under the manifold connections, and once again it is possible to drain the oil continually from these, or to quickly replace them with empty ones, in the case of a persistent leak.

Of no less importance are the tools used for monitoring the flow of oil. If the cargo operations are to be pollution free then very little is going to be seen of the cargo that is being transferred. It must be possible to see the position of valves and to know at what speed the pumps are running. There must be gauges to indicate the pressure in the lines and, of vital importance, others for showing the ullage in the tanks.

The operators

It is said that a good workman will never blame his tools and it is true that the tools are only as good as the hands that use them. Tanker operations are under the control of people and people are, unfortunately, only human and fallible! Human error is a factor in the majority of pollution incidents and the previous pages have shown the areas where this can be most dangerous. Some actual examples are: allowing tanks to overflow, the incorrect setting and control of valves, making an improper hose connection and often no more than a lack of awareness of the hazards involved leads to an accident.

There is a fortune awaiting anyone who can devise a system of training and motivation that eliminates human error in those operations under human control. Nevertheless, it is possible to go a long way towards achieving this as is discussed later in chapter 12 Emphasis has been placed on those areas of tanker operations that involve a risk of pollution, but they are, after all, areas of risk rather than an inevitable source. The number of pollution incidents is minute when compared with the number of operations taking place. It is mainly due to the sound techniques that have been developed over the years and adopted on board all well-run tankers and by all competent tankermen.

JETTY OPERATIONS

All that has so far been said in this chapter about the tanker is also equally relevant when considering the jetty. A similar sequence of events occurs and similar risks of pollution must be guarded against in the same way as on board the ship. Any oil spill from the jetty is just as likely to threaten the marine environment.

As was the case with the ship, the only anti-pollution tools available are the jetty and the equipment on it. Indeed, the principle of containment and the means of achieving it are applicable to oil transfer operations wherever they take place.

The arrangements on the jetty and the operations that are carried out are very much the same as those on the vessel's deck. There are the same hoses which must be connected to manifolds at the jetty end, and the oil flows across the jetty to, or from the shore tanks through the permanent pipeline system. Therefore, the whole working area of the jetty should be surrounded by a raised edge – equivalent to the gunwale bar at the edge of the ship's deck – so as to contain all oil spills. Similarly, under the manifold connections there must be more local containment measures. Also, the whole area must be adequately drained both to clear rainwater and to deal with any large or prolonged spill. Unlike the ship's scuppers, however, these drains can be permanently connected to a slop tank situated beneath the working platform level. As this can only be a tank of small capacity it is essential that the jetty routine ensures it is pumped into a larger shore tank at frequent and regular intervals. To guard against the human error it should in addition be fitted with a high level alarm.

Proper means of communication are equally necessary in preventing pollution. Within terminals the distances between the storage tanks, pumps, and the jetty have always been greater than between equivalent areas on board ships, so there has always been a greater need for immediate communication. This was usually through telephone links with the additional safety measure of remote pump stops on the jetty for use in an emergency. Now, the increasing size of ships has created a similar need for instant communication both between the ship and the jetty and between the different parts of the ship itself. When talking with the jetty the situation is likely to be aggravated by a language problem. It is no longer possible to rely on being able to pass orders and instructions quickly enough by shouting and the complexity of the transfer operations preclude the use of sound signals except as emergency measures. Even a telephone network seldom has sufficient flexibility for present day needs; the only adequate communications system is portable VHF transceivers.

Lighting is another part of the jetty equipment that can play a vital anti-pollution role. Just as most fires start small enough to be easily smothered before they get out of hand, so most oil spills are initially capable of being contained. However, as with fires, unless they are discovered in time and their source eliminated they are likely to quickly overwhelm all defences. So, if transfer operations are to be undertaken during darkness sufficient lighting must be provided so that any oil leak can easily be seen. To satisfy this requirement the area lighted should extend well beyond the immediate working area so as to illuminate the water around the jetty and vessel.

The oil transfers described in this chapter are joint undertakings with no single person being responsible for the whole operation. In general the ship is accountable for all that happens inboard of her rail and the shore installation for everything outboard of it. In practice the situation cannot remain so clear cut; many actions taken on one side of the rail must have consequences on the other — consequences that could easily lead to a risk of pollution.

While it is at the ship's rail that the two areas of responsibility meet there is one vital piece of equipment that must actually link them. A closer look at the cargo hose and its use will serve to illustrate just how accountability merges and responsibility should be shared. Usually it is the jetty that provides the hose or mechanical arms and jetty personnel that make the connections. Therefore, the jetty has the primary duty of ensuring they are in good condition and have been subjected to the statutory examina-

Plate 17: *Aerial view of jetties at BP's Kent refinery.*

Table 9: The sequence of anti-pollution operations in port.

	PREPARATION		TRANSFER		SECURING
	Arr. in port →				→ Sailing
SHIPBOARD PERSONNEL	Planning transfer	Establish communications and emergency procedures	Watchkeeping; checks on deck; hoses/gangway/waterside		Clearing decks; Unberthing
TERMINAL	Check/prepare equipment	Containment measures available; Persons in charge present	Spills contained; Monitor hoses/jetty area		Stow hoses carefully
MOORINGS		Adequate for weather and tidal conditions expected			
SCUPPERS		Plugged	Means available for removing excess water/oil		Cleared; plugs removed
DRIP TRAYS		In position	Means available for removing excess water/oil		Drained
SEA VALVES		Closed and lashed	Ballast?		
HOSES/ARMS		Properly suspended; Coupling tight	Sufficient length		Drained; Disconnected; Blanked
CARGO SYSTEM	Checked	Lined up for transfer			Shut down; Manifold blanked
CARGO TANKS			Ullages monitored		
CHECK LIST		Persons in charge agree to commence transfer			

tions and pressure tests. But at the same time the ship must be prepared to make an inspection before they are taken into use and cannot escape the consequences of accepting defective equipment. Similarly, if the ship's lifting equipment is to be used the roles are reversed. The ship has an initial responsibility to see that it is adequate and safe while the jetty should not accept such assurances blindly. Some terminals ask to inspect the vessel's Factory Acts records before using lifting gear and all should refuse to employ it if a superficial examination suggests there are defects. During the transfer it will be the duty of jetty personnel to actually watch and tend the hose, though it would be foolish for the ship's staff to ignore any dangerous situation that arises. If the consequences are to be shared then the concern must be mutual.

With such divided yet concomitant responsibility it is clear that full cooperation between the ship and jetty should start during the planning stage of any transfer operation when all relevant information should be exchanged. Yet such cooperation should not finish after the initial consultation. The value of adequate means of communication as an anti-pollution tool has already been discussed but it should be remembered that they will only be effective if sufficient and the right kind of information is exchanged.

Just as the ship's rail is the physical point where each area of responsibility merges so the process of marking the 'check-list' immediately before the transfer commences can be seen as the point where a joint accountability is accepted. The presence of both persons in charge is evidence of their mutual concern; their joint signatures an acknowledgement that all the necessary precautions have been taken within their respective spheres. Without the agreement of both the transfer cannot commence, but without their full and continuing cooperation it should not continue.

All of the equipment must be properly designed and regularly tested and inspected both before and during the operations. Table 9 summarises these tasks.

All professionals have a few tricks up their sleeve that make the job look deceptively easy to the amateur; though in reality they are not sleight of hand but techniques learnt only by years of experience and practice. So it is with the prevention of pollution; which means it is time to look more closely at the anti-pollution techniques used on board.

	Precaution/Preparation	Checked
	Name of Terminal: Name of Vessel: Date: Berth: Time:	
1.	Vessel adequately moored for expected weather and tidal conditions.	
2.	Transfer hoses/arms long enough to allow for movement of vessel.	
3.	Each hose supported so as to avoid strain on coupling.	
4.	Cargo system properly lined up.	
5.	Parts of cargo system not in use properly shut down or blanked off.	
6.	Sea suction and overboard valves closed and lashed.	
7.	Hoses in good condition with no visible defects.	
8.	Hose/loading arm connections properly made.	
9.	Containment measures in position.	
10.	Scuppers plugged.	
11.	Means available for draining containment system.	
12.	Emergency shut down methods and procedures available and known.	
13.	Adequate means of communication available.	
14.	A common language is adequately understood by ship and shore.	
15.	Designated and qualified personnel on duty.	
16.	All relevant information exchanged between ship and shore.	
17.	Persons in charge of operations present.	
18.	Adequate lighting available for night operations.	
19.	The persons in charge agree to the transfer.	

Checked by:

For ship: . . . .   For Terminal: . . . .

Table 10: *sample pollution prevention check list.*

ANTI-POLLUTION TECHNIQUES

Every tanker loading or discharge operation can be divided into three distinct stages: preparation, the transfer of oil, and the securing of the cargo system on completion of discharge. *(See table 9)*.

The preparation is just as important in the prevention of pollution as any operation during the actual transfer and the preparatory planning is just as important as any actual preparation. The preparation of a plan and the exercise of committing it to paper will inevitably concentrate the mind onto those operations that involve the greatest risk. Moreover, early and precise knowledge of what is expected to happen will help to ensure that all the necessary resources are at hand and, as the transfer progresses, provide a monitor and checklist. *(See table 10)* It is necessary that all personnel who are going to be involved in the transfer are aware of and fully understand the sequence of events. Not only can they then accept and fulfil a degree of responsibility for preventing pollution, but beyond this, their awareness will provide a further check on the soundness of the procedures adopted. The shore installation must also be included in this planning so as to ensure their co-operation; mutually acceptable procedures must be agreed, areas of responsibility delineated, and adequate communications established.

Also, the vessel herself must be prepared. The need for the scuppers to be plugged has already been mentioned and the possibility of oil escaping onto the deck calls for adequate quantities of absorbent material to be available. Also, if the containment devices are not to be overwhelmed in the event of a large spill means of draining the oil back into the tanks will have to be provided. These consist of drain lines from any fixed drip trays and small, air-driven salvage pumps for clearing the scuppers. The cargo system must be lined up and the position of all valves checked. Finally, but certainly not least in importance, the sea valves and overboard discharge valves must be tightly closed and securely lashed to prevent them being opened accidentally.

The most critical moment of any cargo operation is that when the transfer of oil is actually commenced. It is this time when the effects of a bad joint, a slack nut, or defective hose will first become apparent. For this reason the transition from planning and preparation to the transfer operation itself should always be marked by a thorough check of all the precautions taken and on all the equipment to be used. This is too important for reliance on

	Design criteria	Annual test	Inspection before use	Inspection while in use
Hoses	Oil tightness under pressure eg. Bursting pressure of at least 600lbs per sq inch. Working pressure of at least 150lbs per sq inch.	YES Thorough inspection under static pressure equal to maximum likely to be reached in use.	For external signs of deterioration.	Correctly handled and suspended. No visible deterioration.
Connections	Oil tightness under similar conditions as hoses.	Not applicable	Suitable jointings used. Sufficient bolts used. Bolts not strained. Flanges clean and flat. Bolts fully and equally tightened.	Oil tight.
Containment devices	Adequate capacity. eg. 6 Bbls if hoses 12" or above. 4 Bbls if hoses 6" to 12".	Not applicable	Oil tight. Correctly positioned. Means available for draining.	Not overfilling.
Scupper plugs	Oil tightness. Mechanical if in containment area.	Not applicable	In position.	Remain in position.
Remote operating/indicating equipment, emergency shut downs		YES Proof of effectiveness.	No indications of defects. Emergency stops tested.	No indication of defects.
Pressure gauges	Accuracy within 10%.	YES Proof accuracy maintained.	No indications of defects.	No indication of defects.

Table 11: *Summary of the main equipment concerned with pollution to be found on board a tanker together with their design and testing requirements.*

memory and should always be made against a printed check-list. Moreover, it should be jointly undertaken by the persons in charge on both the ship and jetty. In that way they acknowledge and accept a mutual concern and provide a further check against each other.

The actual transfer should always commence slowly and be maintained at a reduced rate until checks show that all is in order. Throughout the operation there must be sufficient personnel on duty. They should not only concern themselves with monitoring the cargo tanks but also make frequent checks on mooring lines, cargo lines and hoses as well as keeping a careful watch overboard in order to detect any oil leak at an early stage. But they should never forget the special care and concentration necessary when the cargo tanks are reaching the topping-off stage.

Although the risk of pollution is reduced on completion of the transfer it is not eliminated. Unless the flexible hoses are fully drained a considerable amount of oil can flood the deck when the connection is broken. So the scupper plugs should not be removed prematurely. Once the hoses have been disconnected and blanks fitted both to them and to the manifold, drip trays should be emptied and the decks cleaned down. A loaded tanker has a low freeboard and all the precautions taken to prevent pollution in port will be wasted if the first sea shipped carries a full drip tray over the side.

Whilst the need to prevent oil pollution affects tanker operations, even more important is the need for tankers to operate safely. Fortunately there is no conflict between these two vital considerations; indeed, they are complementary. No anti-pollution measure will detract from safety and, similarly, safe operations can only reduce the risk of pollution.

Bunkering operations are really no different from loading a cargo. The preparatory stage is just as important and, as bunkering operations have to be carried out on all ships and are not just confined to tankers, the importance of preparing adequate containment measures needs stressing even more.

Once again it is worthwhile pointing out that all these techniques are just as relevant to jetty operations. The shore installation should have been included in the planning before the transfer commenced.

THE CRUCIAL FACTOR

All of the operations covered in this chapter are also the subject of a multitude of regulations mentioned at the start. These are both national and international; some mandatory and some advisory.

However, good practice and legislation can only go so far, the crucial factor in all anti-pollution measures is the human. In port operations, competent and well motivated personnel can make a bigger contribution towards reducing the risk of pollution than at any other time during the tanker's voyage. There is no easy way of achieving this. It needs a continuing programme of training, supervision and motivation. Above all it needs the development by the individual of an understanding of the grave effects of pollution and the acceptance of his own responsibility to prevent them.

8

The pipelines contribution

E.M. King, FIGasE, MIME, MIMarE, FInstPet
Senior Advisor Pipelines
Gulf Oil Company International

The pipelines which today transport such immense quantities of oil, gas and other commodities are safe, reliable, and environmentally commendable. The minute fraction which is lost due to all forms of spillage illustrates effectively what can be achieved by the application of good engineering principles to competent design and construction, coupled with skilled operation and detailed maintenance.

This chapter describes briefly and simply the basic philosophy of pipeline design, construction, and operation which has resulted in the very high standard of performance attained by this most unobtrusive method of transportation.

DEVELOPMENT

It is generally accepted that introduction of the wheel has been one of man's most beneficial inventions and one that has contributed significantly to his development through the ages. It is known that axle-mounted wooden wheels were in use in Samaria 3,000 to 2,000 years B.C. while it is highly probable that some form of stone wheel was employed even earlier. The chronology of pipeline development may well commence even further back in history and comparison of advances in these two vital forms of transportation makes a fascinating study. Pipes for conveyance of water, sewage, natural gas or brine, and employing hollow logs, stone, lead, bronze and wrought iron can be traced throughout the known history of man. It was probably about 5,000 B.C. that the ancient Chinese first constructed pipelines of hollowed bamboo to carry water from streams and springs to their villages and fields,

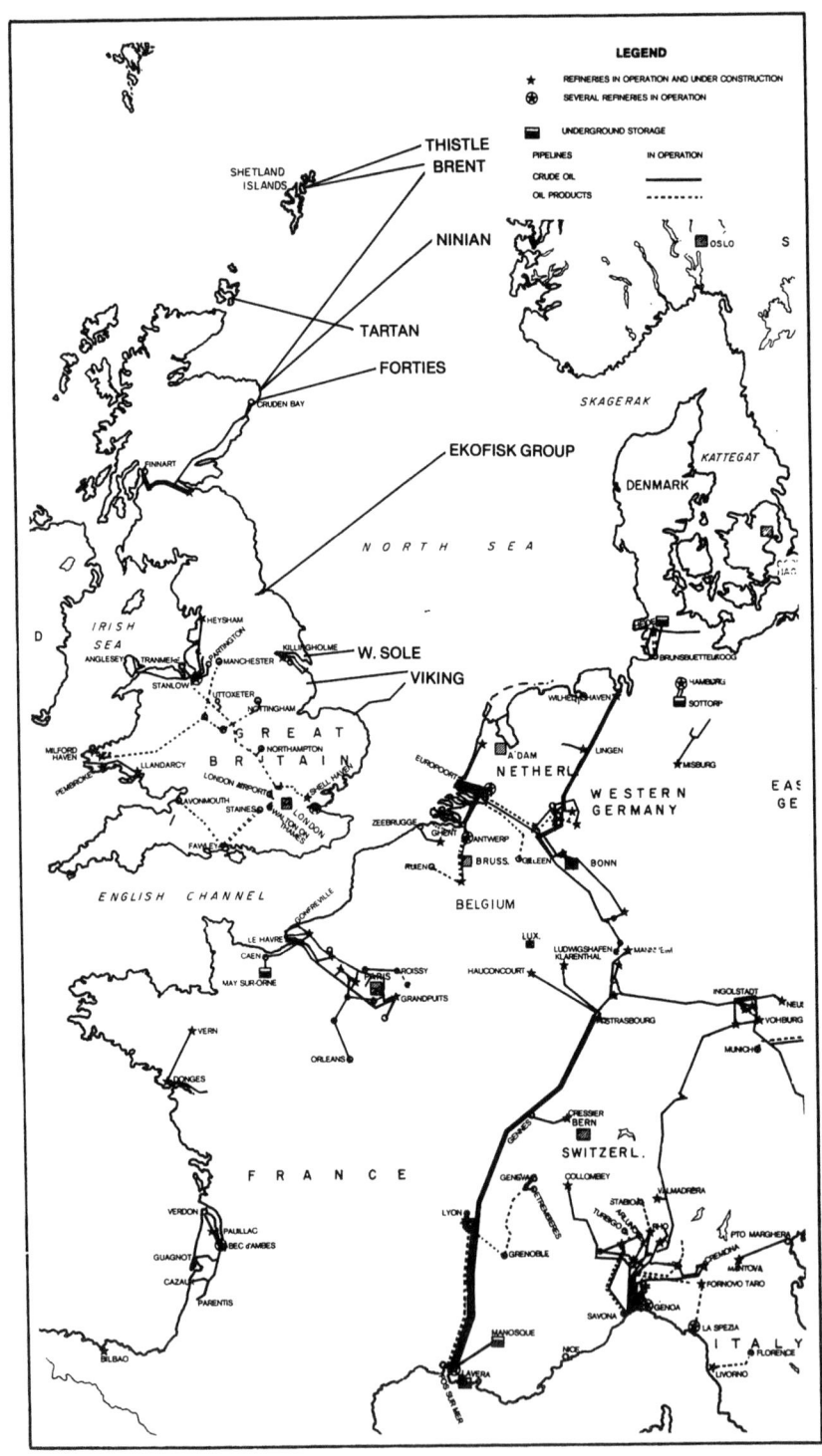

Figure 22: *Major pipelines in Western Europe.*

the stalks being buried and joined with clay or pitch to minimise leakage.

The greatest impetus for pipeline development came from countries where open conduits suffered considerable loss from evaporation and absorption, and in Babylon around 4,000 B.C. water was carried through buried pipes made from fired clay tubes. It is recorded that the Persian army which invaded Egypt through desert territory in 525 B.C., was provisioned with water through a pipeline made from sewn ox hide. The earliest known metal pipes were made from copper, and one specimen found at the Temple of Abussi in Egypt is said to be about 5,000 years old.

As for the oil industry, pipeline development is generally considered to have commenced in the USA in the late 1860s with small bore cast iron pipes for gathering of crude oil from production wells and transportation to loading railheads. One such line was that laid between a producing field at Pithole City to a loading point on the Oil Creek Railroad in Pennsylvania. This line was 2 inches nominal bore with a daily capacity of about 250 tons. The first long distance line is believed to be one laid in the late 1870s over 100 miles in length and carrying crude oil over the Allegheny Mountains. Thenceforward crude oil lines began to proliferate in the USA and with steel starting to replace cast iron, some 40,000 miles of line were in service by 1910. The first recorded modern counterpart was probably the all-welded steel pipeline constructed in 1924 in 14, 16 and 18 inch diameters to convey well gas between Louisiana and Texas. Very rapid development of high pressure steel pipelines followed, with the availability of seamless pipe in 1925, the replacement of oxy-acetylene welding by electric arc in 1928, the introduction of cathodic protection to combat corrosion in 1930, and the introduction of non-destructive testing techniques with X-ray inspection of welds in 1946.

The USA has long been acknowledged as the pioneer of present day pipeline construction techniques, which except for increases in dimension and improvement in mechanical plant, have followed a similar general pattern for the past three or four decades. These techniques have been adopted and adapted to suit local environment and geography the world over, but the basic procedures vary only in minor detail.

For reasons of physical strength, high grade steel remains the universal medium for oil pipeline construction, and buried land lines now exist in diameters up to 48 inches, with up to 36 inches for long distance submarine pipelines, although short sections up to 56 inch diameter have been laid to offshore tanker loading facilities. Most of the new generation lay-barges being built for pipe-laying operations in the North Sea are capable of handling and installing up to 52 inch diameter pipes.

PERFORMANCE

This book deals specifically with prevention of pollution from oil, therefore if spillage of oil can be avoided, pollution will not be a problem. Unless some extraneous agency interferes with its safe operation, an oil pipeline designed and constructed to modern standards and operated in accordance with accepted controls and practices has an extremely low potential for loss of any fluid content. Further, if a spillage does occur, the probability is that the effectiveness of present day recovery techniques will result in most of the product being recovered or safely disposed of with no lasting harmful effect to the environment being sustained.

Statistical data collated by such organisations as Stichting CONCAWE[1] in respect of Western Europe, the Department of Transportation's Office of Pipeline Safety for the US[2] and the Canadian Petroleum Association,[3] emphasises the inherent safety of oil transportation by pipeline and each of these organisations reports factually on the incidence, volume and causes of leakage from oil industry pipeline systems.

International airlines emphasise the relative safety of modern air travel by a comparison of mishaps to passenger-miles flown. Similarly, the safety record of the oil pipeline industry far exceeds that of any other mode of transportation on a volume-distance basis. During the past ten years, oil industry pipelines in Western Europe have averaged one spillage incident for every 594 million cubic-metre-kilometres, and the total volume spilled has amounted to less than 0.0004 per cent of the volume carried, or four parts per million. The bulk of even this insignificant amount has been recovered at site and the resultant effect on the environment has been infinitesimal indeed. The major oil industry pipelines in Western Europe are shown in figure 22.

Pipelines are the only form of bulk trans-shipment which effectively reduce surface congestion of roads and waterways, and they are so secure, reliable and unobtrusive that the general public is normally totally unaware of the vital arteries of energy beneath their very feet. Nobody is more conscious of the effects of environmental pollution than those responsible for maintenance of potable water supplies for human consumption, and the International Waterworks Association, composed of water supply authorities from some 67 countries, has agreed that pollution from oil pipelines is the least hazardous of all methods of transportation currently in use.

DESIGN AND CONSTRUCTION

Clearly defined national or international standards and codes of practice are available for every aspect of present day pipeline design and construction, although mandatory requirements differ from country to country, and not all states have their own indigenous codes. The standards adopted have been developed over a considerable period to an exceedingly high standard. They are kept under constant review, and are extended or amended whenever there is room for improvement. Ideally, these or similar standards and codes should be applied to all pipelines, no matter where they may be built.

In the UK for instance the construction of pipelines is regulated by Acts of Parliament, passed in 1962[4] for underground and in 1975[5] for submarine pipelines. The British Standards Institute has issued standards on the design and construction requirements for landward pipes[6] and for field welding of carbon steel pipelines.[7] The Institute of Petroleum has issued a code of practice for petroleum pipelines.[8] Similar standards and codes have been issued in the USA.[9]

The design of a pipeline or pipelines system is governed predominantly by the physical parameters pertaining to its required operation. For instance, the pipe diameter, grade of steel, pipe wall thickness, control and block valve locations will be largely determined by the type and volume of product to be moved, the pumping pressures required to move it, variations in pressure due to changes in elevation, and geographical characteristics of the route. However, an increasing number of design considerations now relate more to safety (i.e. the physical integrity of the system and retention of its content) than to functional parameters. This applies particularly to submarine installations from lay-barges or pulled estuarial crossings, where the grade of steel and wall thickness is more likely to be governed by the bending and tensional stresses imposed during installation, or resistance to environmental stresses due to depth of water, bottom currents, etc.

One of the first prerequisites to pipeline construction is the establishment of a safe and acceptable route between the terminal points. An environment which provides maximum security for the pipeline will obviously minimise possiblity of pollution of that environment by the contents of the line. Careful regard must be paid to excessive changes in elevation which will affect fluid head, pipe wall-thickness, valve locations, drain-down volumes, or necessitate installation of pressure reducing equipment. Consideration must be paid to ready and rapid accessibility to the pipeline right of-way for routine patrolling and maintenance purposes, and by emergency vehicles and equipment. Areas scheduled for future development or new road construction would be avoided as far as possible. Rivers, lakes, water catchment areas, locations prone to flooding or subsidence would be circumvented, or special measures adopted if no acceptable alternative existed. Careful investigation will usually identify areas of current or past mining activity and the possibility of future subsidence. Mining authorities are normally extremely co-operative and will provide details, when available, of their subterranean activities and will frequently volunteer predictions as to where and when surface movement may become evident and to what extent. No selection of route would be considered complete without a comprehensive soil resistivity survey to identify any potential areas with particularly aggressive soil conditions, to determine the number of cathodic protection application points required to inhibit corrosion, and the most suitable locations for these installations.

The planning of routes for submarine pipelines may be considered to be simplified by the probability that a straight 'line-of-sight' can be followed far more frequently than is possible on land. Nevertheless a number of special surveys are usually necessary to confirm the nature and acceptability of bottom conditions. These would include an echo sounding survey to provide a pattern of soundings from which the bathymetry of the seabed could be determined and a comprehensive contour plan produced, a seismic profile survey to locate and determine nature, extent and levels of various strata for pipe bedding and burial purposes; a side-scan sonar survey to detect bottom irregularities, outcropping rocks, and other possible protrusions standing clear of the bed; and a manual inspection carried out by divers in workable depths, or by manned or unmanned submersible craft equipped with cameras and recording equipment. These surveys would be supplemented by measurement of tidal and current effects, weather data, etc.

A detailed description of pipeline construction techniques would diverge from the entitled objective. However, certain procedures specifically concerned with the integrity of the pipeline and its contents deserve mention. As one of the prime components involved, the line-pipe itself is subjected to rigid quality control from initiation of a project, until the final acceptance tests of the completed facility. The grade of steel is selected according to the duty for which the pipe is destined, its required performance in operation, the environment in which it will be installed, and the stresses to which the fabric will be subjected both during installation and for its designed operational life. The pipe may be required to carry a multiplicity of different products, or it may have to be operated in a cryogenic state or at above ambient temperatures. The required operating pressure coupled with the geographical profile and related to the specified yield stress of the steel, will together determine the wall thickness for the particular pipes in each location. In the final analysis a range of differing wall thicknesses may result throughout the overall length of any one pipeline. Other influencing stresses likely to occur during construction or operation may necessitate further increases in pipe wall thickness, as for instance those imposed during marine pulling or submarine lay-barge installation. For any given pressure rating there will be various combinations of steel grade and wall-thickness which may be considered. The higher stress steels now available enable a much thinner pipe wall to be employed for the same ultimate pressure containment. However, there are occasions when a more rigid heavy-wall pipe is preferable, for example to minimise deflection or distortion, or perhaps for extra protection through a built-up area where proliferation of service excavations can increase the risk of damage to a thin-walled pipe.

Steel pipe in the various grades may be produced as a solid-drawn seamless tube, or with electrically welded longitudinal or spiral seams. During manufacture the chemical analysis of the steel, as well as fabrication procedures, are controlled to extremely fine tolerances, and each process is subjected to close inspection. Each pipe undergoes an individual hydrostatic strength test on completion and ultra-sonic or other non-destructive tests may be applied. Selected samples undergo separate destructive tests for longitudinal tensile strength, transverse tensile strength, yield stress, flattening, fracture toughness, elongation, bending, weld ducticity, and weld tensile strength. The American Petroleum Institute specification referred to above for the manufacture of steel pipelines is most widely employed in the oil industry. Only carefully selected and approved mills are authorised to manufacture and classify pipe as being to API standard.

Having produced a pipe to such a high degree of mechanical proficiency, it is necessary to ensure that the method of joining individual lengths together will provide comparable characteristics of quality. Although flanged or other mechanical joints may be used above ground, or in other accessible locations, for insertion of valves and connection to various appurtenances, electric arc welding is now the only acceptable method for coupling buried steel pipelines. Welding is usually totally manual, although semi-automatic, or fully automatic techniques are available for certain operations. Experiments are also in progress for developing a satisfactory friction-welding technique for field use with pipelines.

For electric arc welding, the 'profile' at each pipe end is machined to specific tolerances; welding rods are selected according to the grade of steel employed, the wall thickness involved and the particular welding procedure specified; and the number of 'runs' to execute a complete weld are determined. Only certified welders are employed for pipeline work, each has to undergo periodic requalification, and these craftsmen are acknowledged to be amongst the most skilled in their profession. Nevertheless, welds are still subjected to rigorous non-destructive testing by X-ray or gamma radiography, by ultrasonic examination, magnetic particle or dye-penetrant detection, or any combination of these. As a further safeguard to ensure preservation of consistent quality, a weld is occasionally selected at random, cut from the pipeline, and subjected to the full range of destructive tests for tensile strength, ductility, and other necessary characteristics.

All other components and equipment associated with a pipeline system are subject to the same high quality of design, manufacture and testing as that adopted for line-pipe.

To guard against inflicting stresses on any part of a pipeline system in excess of the predetermined safe working limits, it may be necessary to incorporate various mechanical safeguards in the original design concept. For instance, in mountainous regions it may be necessary to ensure that the specified pressure maxima are never exceeded by installing a series of pressure relieving or pressure reducing stations; pressure relief or alternative safety equipment would be provided if surge generation were likely to exceed the safe operating limits; and provision for thermal pressure relief would be installed where ambient temperature rises could affect locked-in sections of pipework. Construction of pipelines systems to operate under cryogenic conditions or at high temperatures each call for particular safeguards to prevent failure in operation. Both must be thermally insulated, of course, for maintenance of temperature levels and to avoid adverse interference with the environment such as the destruction of beneficial soil bacteria, reduction of moisture levels, or interference with crop propagation in agricultural land, and the necessity to avoid injurious temperature effects in regions of permafrost. The cryogenic system will require specific provision for contraction which may otherwise induce intolerable stress in components, may affect leak-tightness of joints, or may induce brittle fracture if unsuitable metals are employed. According to the particular duty and operation a 'hot' line for pumping very viscous oils may be designed to allow unrestrained movement to counter the expansive forces, or alternatively all stresses may be locked-in and the system totally restrained. The latter technique is frequently employed for pipelines systems and calls for a more carefully analysed design.

Every phase of pipe preparation, wrapping, welding and installation is subject to inspection and supervision by highly qualified and experienced personnel, and the strict standard of testing imposed on progressive stages provides that the completed system shall be virtually flawless. Even the pipe trench is carefully cut and an even bed prepared to avoid damage to the wrapping, or local stresses due to spanning between high points. Similarly only selected fine material is used for the initial back-fill around the sides and over the top of the pipe. Correct compaction of the fill immediately adjacent to the pipe prevents the disturbed channel becoming a natural land drain with possible scouring of the protective wrapping, and pockets of water and air forming aerobic conditions for the promotion of corrosion, and in rocky areas sand or other acceptable fill is imported and used for trench levelling and lining before the coarse spoil is returned and compacted.

The final phase for modern pipeline construction is commissioning of the cathodic protection system, checking that the level of protection is adequate and coverage complete, and that there is no adverse effect on any existing foreign structures.

An outline of pipeline construction techniques would be incomplete without mention of the rapidly expanding network of submarine pipelines, particularly those in such hostile environments as the North Sea, with their associated problems of design, construction and maintenance.

In the sphere of underwater pipelining, installation techniques may be divided into four main categories. The first two have been used in one form or another for many years, mainly in respect of comparatively short sections across rivers, lakes and estuaries or for offshore services, and maybe referred to briefly as a 'flotation' method, whereby the complete pipeline section is fabricated on land, launched with temporary buoyancy floats, or lifted by floating derricks, manoeuvred into precise position over the excavated trench and sunk into place; the 'pulled' method, where the pipe is welded up on shore into sections, or strings, and is subsequently winched into a prepared trench, or is pulled out onto the sea floor and buried by jetting plough. The third process which is used for offshore lines too long to be pulled is the 'lay-barge' method of construction, whereby all the functions of a land 'spread' are condensed into a number of sequential processing stations on a floating, completely self-contained construction unit. Pipes already prepared with insulating wrapping and concrete weight-coat are delivered to the lay-barge where they are welded into a continuous length, fed over an underwater 'stinger', an articulated support frame to control curvature, onto the ocean floor and may be eventually secured in position by underwater anchors or buried. The 'first generation' lay-barges originally equipped for operation in fairly shallow waters in the Gulf of Mexico were merely massive pontoons with all the necessary equipment mounted thereon. With the advent of deep water pipelaying, similar but much larger craft were built. This pontoon type of vessel has its operation severely curtailed by bad weather, which entails the end of the pipe being capped, buoyed and laid on the sea bed under carefully controlled conditions, while the barge goes off station temporarily and returns later to recover the pipe end and resume laying as weather conditions improve. This extremely expensive process has been largely responsible for the 'third generation' lay-barges now employed, which are either of orthodox vessel hull type, or semi-submersible, with underwater hulls supporting a construction deck on columns above water level. *(see plate 18)* Both types of vessel are fully self-manoeuvrable and able to continue operations in far heavier sea conditions than was previously possible. Further developments of the lay-barge type of construction parallel the two first described methods and are known as 'surface tow' or 'bottom tow' techniques. Long lengths of land prefabricated pipe are towed out to site either floating at the surface, or by carefully controlled buoyancy just above the sea bed, and jointed into continuous

Plate 18: *Semi-submersible lay-barge* Choctaw II.

lengths by a specially adapted vessel similar to a lay-barge. The fourth category has very restricted application and may be used only for smaller diameter sea-lines. A vessel known as a reel barge is able to accept a continuous length of shore fabricated pipe onto a massive rotating drum. The barge proceeds to location, joins the new length to previously laid lengths and reels out its new load onto the ocean floor, passing the pipe through a straightening device to remove drum induced curvature. Prototypes of this method were employed as far back as 1944 when the 'PLUTO' cross channel lines were laid to supply allied invasion forces with petroleum products. The 2.5 inch diameter 'HAMEL' type pipes were reeled off from a 32 feet diameter floating drum, known then as a 'Conundrum' because of its conical ends; and the 5 inch diameter 'HAIS' type lines were reeled out from a submarine-cable laying vessel. Present day reel barges have laid up to 12 inch diameter pipe, a vessel is under construction which will be able to lay 9 kilometres of 16 inch pipe at each loading, and designs exist for possible future construction which will handle up to 24 inch diameter.

Pipelines up to 36 inches diameter have been laid in up to 170 metres of water by stinger-equipped lay-barges; 12 inch pipe has been laid at 300 metres depth in the Gulf of Mexico by reel barge; 10 inch in Lake Geneva by surface-tow method to 340 metres depth; and 16 inch has been laid experimentally in the Mediterranean by lay-barge and stinger to a record depth of 570 metres.

OPERATION AND CONTROL

The primary objective of the pipeline operator is to transport safely a given volume of product from a reception point to a delivery point without loss or contamination. In the early days of pipeline operation, the most sophisticated aid to safe operation and control of related activities was a telephone or R/T radio. Many years ago, a pipe route across remote unfarmed areas was frequently identifiable by the pair of pole mounted wires strung between intermediate pump stations. Both forms of communication still form an important link in pipeline operations, but utilisation of a voice channel is an auxiliary function only. The modern pipeline system still has its main control centre, normally manned twenty-four hours a day, each day of the year, but frequently only by a single operative. Process control, monitoring of status, activation and response to alarm conditions, may all be performed tirelessly, accurately and with infinite repeatability by supervisory systems masterminded by mini — or micro — computers, which are linked to local mini-processors at each outstation. The master unit sequentially interrogates each remote unit, receiving back all relevant data pertaining to that outstation, e.g. pressures, temperatures, tank levels, meter readings, valve conditions, pumps in operation, flow conditions, etc. On instruction from the master controller, the local unit will perform various operational functions such as opening or closing valves, starting or stopping pumps, initiating prover runs for checking meter calibration, etc. On a priority over-ride basis, the local unit will detect an alarm condition and through the master unit will initiate appropriate action depending on the malfunction. Any defect which creates or could lead to a hazardous situation will automatically shutdown the system in a safe and controlled manner. The operator at the control centre is able to identify the source of the alarm and to receive information on the cause. The system can only be restarted manually, but the supervisory control system will inhibit reactivation until the alarm condition has been remedied. The essence of such a system of control is the speed of process and application which cannot possibly be matched by the human element. A master unit can interrogate several outstations, receive and process data from each of them, call for verification if necessary, recheck and display all the relevant information in only a few seconds. Thus the time lags associated with human reaction, communication, calculation, decision and implementation are reduced to virtually immediate detection and remedial action with the modern pipeline supervisory systems.

CAUSES OF SPILLAGE

The aspect of pipelines constructed to present day standards and with modern systems of control has been emphasised, but this in no way detracts from the excellent record of performance and integrity relating to the many hundreds of miles of older pipeline networks, which is adequately illustrated by the published statistical reports on oil losses from pipelines.

In spite of all the efforts of the designers, constructors and operators, spillages do occasionally occur from oil pipelines and it

is suggested that these may be classified in five main categories:
1. Mechanical failure
2. Operational error
3. Natural hazard
4. Corrosion
5. Third party activity

Mechanical failure

This cause may be further sub-divided into failures attributable to faults in construction and those arising from defects in materials or components. While historically both sources have been responsible for spillages from pipelines in the past, the very high standards of construction and inspection applied by the oil industry have drastically reduced the potential for this type of failure.

Operational error

Again, this category may be split into two different classes, namely system malfunction and human error.

Some older pipelines are still operated on a predominantly manual basis, and all pipelines have some manual functions incorporated into their operation, even if only during maintenance activities. All essential functions are normally protected by interlocks and safety overrides, even for manual operation, and analysis of data relating to this type of spillage indicates that any mishap resulting from human error would normally affect only an enclosed terminal or pump-station and only a very minor volume would be involved. Incidents recorded during recent years have included such errors as leaving a small vent cock or drain tap open after maintenance works, and not fully closing a sampling point.

System malfunctions resulting in spillage have been virtually eliminated during the past decade thanks to the protective interlocks already mentioned and the fact that the modern pipeline systems are largely processor controlled with built-in fail-safe devices to safeguard every operational function. Normally the worst that can happen on such a system if any particular component or item of equipment fails to operate as designed is that the pipeline will safely shut down, and cannot be reactivated until the fault, or apparent fault, has been identified and corrected.

Natural hazard

This relates to damage arising from landslides, subsidences and flooding. Such occurrences are in any case extremely rare, and the risk is reduced by careful routeing investigation to avoid possible problem areas such as occur in mining regions, and by intelligent observation and interpretation of warning signs during periodic inspections and patrols. The few recorded incidents that have caused spillage from pipelines have generally resulted from very sudden and unpredictable freak weather conditions.

Corrosion

Every metal has a natural tendency to revert back to its original mineral form, and the more highly processed the finished metal, the more easily the reversion phenomena occurs.

Corrosion of steel is an electro-chemical reaction between the metal and its environment, in which the steel reverts to iron oxide. As far as oil pipelines are concerned, thanks largely to the nature of the commodity itself, internal corrosion of the pipe wall is rarely a problem, but it has occurred on facilities such as jetty loading lines which have been flushed with brackish or sea water and incompletely drained, on lines which have been subject to only intermittent use, or on gathering systems for crude oil which has a high water and dissolved oxygen content.

The outer wall of a buried steel pipeline is most at risk from corrosion, which can usually be attributed to one, or a combination of, those prime conditions which initiate metallic deterioration. These may be classified briefly as follows:

Galvanic corrosion
When two dissimilar metals are connected together in an electrolyte a current will flow from one to the other. The metal from which the current flows (anode) will tend to dissolve in the electrolyte, leaving that receiving the current (cathode) not obviously affected.

In a buried pipeline, the currents in question may be generated between areas of different composition in the metal itself, or between areas of pipe surface in contact with environments of different composition. Differences in oxygen concentrations at separate areas of pipe surface may also produce a galvanic cell, and the conductivity of the electrolyte, i.e. the surrounding medium which completes the circuit between anodic and cathodic areas, generally regulates the flow of current. Thus, a pipeline lying in wet clay soil of low resistivity, may be expected to be at greater risk than one in a dry sandy environment with high resistivity.

Stray current electrolysis
Rapid and severe corrosion may occur if stray D.C. currents are short circuited onto a buried pipeline in one location and returned to earth at another point. Such occurrences have resulted from faulty connections to electric railway or tramway motive power circuits, from arc welding installations, or from incorrect earthing of appliances to service pipes. Awareness of this phenomenon has resulted in precautionary measures which have largely eliminated this source of corrosion.

Bacterial corrosion
Many interesting, lengthy, and involved articles have been written by specialists on the subject of bacterial corrosion, but for the reader of this publication, a brief and simplified explanation should suffice for the problem to be appreciated.

The normal corrosion reaction from moisture in contact with steel produces a coating of iron oxide (rust) on the wall of the pipe, together with a cathodic barrier of hydrogen ions which tends to inhibit further corrosion. Absorption of the hydrogen by soil bacteria will not only allow the corrosion process to continue, but can materially accelerate the reaction.

Bacterial reproduction is normally effected by consumption of oxygen, but one type called the desulfovibrio group proliferate by reduction of sulphate in the soil. Exposure to oxygen will destroy this bacterial form and consequently risk of corrosion from this source is generally confined to anaerobic areas in well compacted strata, particularly those plentiful in sulphates and other soluble salts. Such concentrations generally exist in the blue and, to a lesser extent, the yellow clays. The hydrogen generated by the primary oxidation process is converted to hydrogen sulphide by the sulphate reducing bacteria, and the corrosion cycle proceeds. It is the hydrogen sulphide which is responsible for the characteristic smell when active colonies are exposed.

Of all causes of leakage, corrosion is probably the least understood by the layman, and is generally considered by him to be the most prolific contributor to pipe perforation. The fact is that nowadays comparatively few corrosion spillages occur, and those that do, generally result in only a minute loss of product before detection. It must be emphasised that corrosion of steel is a gradual process of wastage and the first, and frequently the only, escape of product from a corroded pipe is a mere drip. This will, of course, gradually increase in volume if corrosion is allowed to develop unchecked.

Third party activity

This category of spillage causes from pipelines is self-explanatory and may be broadly sub-divided into three groups. The most predominant comprises incidents which arise from physical damage inflicted on a pipeline, usually by heavy excavation, earth moving, or other mechanical plant, and is attributable to non-intentional injury, frequently referred to erroneously as 'accidental' damage. Often, such incidents arise from lack of communication, carelessness, negligence, or failure to observe recognised safeguards, as well as, occasionally, true accidents.

The second group includes those occurrences where damage is suffered indirectly as a result of another party's work executed adjacent to a pipeline. Again, this is invariably unintentional, although occasionally it may be attributable to the adoption of unacceptable engineering practices. For instance, lines have been affected by minor landslides initiated by the collapse of unsupported deep trenching excavated in unstable ground close to a pipeline route; a pipeline river crossing was recently disturbed by a flash flood following a violent storm due to the fact that undisclosed gravel recovery from the river bed had reduced the cover over the pipeline to a critical level; other similar cases may be cited.

The third type is extremely rare but has occurred in the past and must receive mention. This involves malicious damage or deliberate vandalism. Surprisingly, although the potential for premeditated damage both to a pipeline and its contents and to the environment is immense, with perhaps a single exception, such incidents as have occurred in recent years have had little impact on spillage statistics, either as regards number of incidents or volume of product spilled, and they have been of nuisance value only to the operators concerned.

METHODS OF DETECTION

In the earliest days of oil transportation by pipeline, only very elementary checks were possible to compare quantities put in at one end and those delivered at the other. Reliance was generally based on tank 'dips' i.e. physically measuring the depth of fluid in a tank and obtaining its volume from a calibration chart. Any large differential was soon identified, but the only method of locating small leaks relied on detection by foot patrols walking the pipe route. Similar patrols are still carried out periodically today, not to look for leaks, but to assist in preventing them by keeping observation on all development and excavation works likely to encroach on the pipe track.

Before oil is first introduced into a pipeline, a hydrostatic pressure test is imposed to check leak-tightness and physical strength, usually with clean water as the pressurising medium. The advent of the all-welded steel line enabled the standard of these tests to be such that there is no allowable tolerance and the test pressure must be held for a stipulated period, usually 24 hours, without detectable loss before being accepted for service. In the U.K. the Governmental Pipelines Inspector requires proof that such tests have been satisfactorily performed and may witness their execution if he so wishes. After commencement of operational use, periodic hydrostatic retests are still a mandatory requirement for checking leak-tightness in some countries. These are usually carried out at maximum operating pressure with the normal line content. Tests may be carried out either over the total length of pipeline, or by sectionalising with special block valves to ensure total segregation of each section. A more recent development is differential pressure testing whereby adjacent sections of pipe are balanced against each other. Only a minute loss is necessary from one section to generate a swiftly detectable differential gradient.

The very rapid development of flow meters for pipeline use and their extreme accuracy, coupled with micro-computer processors for data assimilation and alarm enunciation, nowadays enables the integrity of a pipeline to be monitored continuously and appropriate action initiated by the supervisory equipment as soon as any discrepancy is detected. Present day turbine and positive-displacement meters have to meet standards of accuracy within 0.15 per cent linearity and 0.02 per cent repeatability in order to qualify for approval by Customs authorities for custody transfer purposes, and specifically designed metering systems for monitoring line integrity frequently exceed even these close tolerances.

A further safeguard for the detection of sudden large spillages, such as would occur if a mechanical excavator ruptured a pipeline, is achieved by the use of flow-rate meters as opposed to volumetric meters. The meters are installed at either extremity of a pipeline section and their output is fed back to control equipment where the differential balance is continuously monitored. Any departure from the preset limits of differential will initiate an alarm and automatic shut-down. Under steady flow conditions, flow-rate is in balance along the system. A sudden increase in flow in an upstream meter and a corresponding decrease in the downstream will indicate a line-break condition between the two meters and system shut-down will be commenced immediately.

For metering accuracy to achieve the required limits, constant correction is applied in respect of temperature, gravity, and viscosity of the product in transit, and a fully comprehensive system will include built-in meter provers for periodic recalibration.

A relic of the early pipeline days is the 'pig' or 'go-devil' which was originally a product-propelled mechanical device inserted into the flowing stream for cleaning the interior of the pipe and removing wax formations. This device carried sprung steel scrapers or wire brushes and frequently included a mechanical noise maker to assist in following its passage along the pipe from valve chamber to valve chamber. The name probably derived from the metallic squeal which accompanied its travel. Cleaning pigs are not now in frequent use, although similar devices are used for evacuating a line (swabbing pig), checking the diameter (gauging pig), or for separation of different products (batching pig). Both swabbing and batching are more usually performed now by neoprene spheres, which may also be equipped with a radio isotope for location purposes. A new porcine family has now been bred for the benefit of the pipeline world — the 'intelligent pig'. These are most sophisticated animals who play an important role in leak detection and prevention, and frequently comprise several articulated sections, known as a 'train', and capable of negotiating the various bends in the pipework. Most possess electronic equipment, distance recording apparatus, self-contained power packs, location devices, and other necessary appertunances.

One intelligent pig very accurately calipers the internal circumference of the pipe as it travels and records all indentations and irregularities in the pipe wall. Another induces an electro-magnetic field into the pipe wall as it travels and records all changes in pipe wall thickness and can indicate corrosion craters. This apparatus is so sensitive that it is possible to identify the position of welds when the recording tape is processed and interpreted. A third type operates on an acoustic basis and is able to record the vibrations produced by product leaking out through a small perforation. A fourth technique employs a radio-active tracer, of very limited intensity and life, which is injected into the product stream. A minute quantity of this tracer would be forced out through any existing leak into the soil adjacent to the pipe. After flushing with unmarked product, a detector pig is run along the pipe and this will record and position any radio-active material leaked from the line.

There are other means of leak detection which may be suitable for specific locations. If a suspected source of leakage has been narrowed to a fairly short length of pipeline under an unmade surface, depending on the product being pumped, it may be possible to inject nitrous oxide into the flowing stream and use manually operated detection probes to locate traces of the gas leaked into the soil.

Where sectional pressurisation is not practicable due to absence of valves in particular locations, a sphere may be inserted into the line and propelled to the vicinity of the suspected leak. The system is shut down with pressure equalised on either side of the sphere and its movement monitored. Leakage will cause a reduction of pressure on one side of the sphere and the higher pressure on the sound side will cause movement towards the leak. Traverse of the sphere down the line and repeated movement checks will locate the source of the leakage.

Many ingenious attempts have been made to produce a viable means of leakage detection external to a pipeline. One such method was a soluble cable laid adjacent to the pipe which, in contact with leaked product, would break circuit and initiate an alarm. Obvious difficulties included the possibility of material other than leakage from the pipeline affecting the cable; the self-destruction of the system in operation; and the difficulty in locating precisely and quickly any source of trouble. Another system, which is available, involves detection equipment placed at intervals along a pipeline to measure negative pressure waves which are generated under line-break conditions. This apparatus can provide very accurate location depending on the number and spacing of measurement points, but will not detect small, steady leakages.

Although most spillages would be more readily and speedily detected by other means, the actual location of the source of spillage on a long cross-country pipeline may be identified by infra-red aerial photography.

Development work is still proceeding on new and improved methods of on-stream leak detection, but the oil pipeline industry in general accepts that prevention is better than cure.

PREVENTION OF SPILLAGE

Apart from the normal pride in achievement derived by those responsible for any mechanical engineering function, it is suggested that there are six dominant incentives for safe operation and avoidance of spillage from an oil pipeline. Without attempting to assign priority, these may be listed as follows:-

Loss of product;
Loss of operating time;
Cost of repairs to pipeline/equipment;
Cost of damage to property;
Effects of pollution on environment;
Effects of adverse publicity on company/industry image.

The dependance of a leakage-free pipeline system on good design and well engineered construction, coupled always with efficient inspection and experienced supervision, cannot be over-emphasised.

However, as has been discussed earlier, spillages can still occur, and measures for prevention may be considered as they relate to each of the five tabulated causes.

Mechanical failure

The influence of design and construction quality is perhaps most effective in this sphere. Design includes specification of materials and components, determining applicable standards for construction, and detailing parameters for inspection and testing.

The aspect of mechanical soundness has been dealt with earlier at some length and although it must be acknowledged that failures in any mechanical components can never be totally eliminated, the very high standards of design and construction adopted by the industry have reduced the potential risk to an absolute minimum as far as oil pipelines are concerned.

Operational error

On statistical evidence, oil spillages resulting from human error have been a very rare occurrence, and as outlined in earlier comments, such incidents as have occurred have generally been of a minor nature within the confines of a terminal or pump station, and have usually been associated with maintenance works rather than routine operation.

Similarly, spillages arising from system malfunction no longer present a serious hazard due largely to the various fail-safe protective devices and system interlocks incorporated into control and processor equipment. Any malfunction will normally enunciate an alarm condition, and if appropriate will activate a fail-safe sequence. If the malfunction is even remotely likely to produce, or lead to, a hazardous situation, the system will safely shut-down operation and will inhibit any attempt to restart until the fault, or potential fault, has been established, located, and rectified.

Natural hazard

Freak weather conditions and geophysical phenomena cannot be controlled, but intelligent anticipation, based largely on past experience, can provide a means of avoiding or reducing damage to a pipeline from such occurrences.

With regard to ground subsidence, the necessity for careful route investigation and selection, and the need to avoid areas of potential surface movement, has been emphasised already. There are occasions when total circumvention of mining regions is impracticable, and if sufficient data is available, adequate precautionary measures may justify acceptance of the route. One such procedure which has been employed, but is now considered less preferable to alternative measures, was to 'snake' the pipeline through areas of possible subsidence, so that slack was available to provide a degree of flexibility. In some cases a form of flexible joint was utilised, but this usually entailed some system of constraint to prevent distortion in the event of movement. The more usual precaution is to install a number of reference stations, from which relative movement of the pipeline or its environment may be gauged. This requires continuous and consistent checking. If and when movement is detected, or is anticipated, a series of strain gauges may be attached to the pipe wall and connected to a remote indicator and alarm system.

Potential landslide locations and areas where freak conditions could produce flooding are often identifiable, and if unavoidable, suitable precautions may be taken during installation. In the latter respect special ground anchors, saddle type trench weights, or a reinforced concrete jacket may be employed to prevent any tendency to float under fluidised soil conditions.

Efficient patrolling of existing lines by foot or by air, can sometimes identify a developing hazard from natural sources, and extra vigilance may be justified during, or subsequent to, prolonged abnormal weather conditions.

Manned, and unmanned submersible craft are used for routine patrolling and inspection of submarine pipelines, and one particular hazard has been revealed and countered as a result. This phenomenon which has affected lines laid on the seabed arises from currents normal to the pipe track causing vortex shedding on the downstream side of the pipe. Sand is eroded from that side of the pipe until the depth of scour produces a void below the pipe invert. The current will deepen and widen this gap until a considerable length of pipe is left unsupported. Such 'spanning' may be countered by burying the pipe, if this is in fact practicable in the depth of water concerned and the bed strata will permit; by providing artificial cover with heavy granular material or crushed rock; by purpose-made concrete covers; or by a more recently developed technique whereby bunches of buoyant stranded polypropylene are anchored in the form of a mat on either side of the pipe. The resultant floating curtain effectively reduces water velocity causing suspended particles of sand and silt to drop to the bed and gradually form an artificial sandbank. The pipe is eventually covered by reversal of the hazardous erosion process.

Corrosion

It has been stated already that, due in some measure to the nature of the products pumped, internal corrosion has proven to be responsible for very few leakage incidents on pipelines. Pursuing the theme that prevention is preferable to cure, care is now taken to minimise the deliberate introduction of water into an operating pipeline, unless an emergency situation warrants its use. Marine service lines in intermittent operation which used to be purged with sea water prior to draining are now frequently cleared by means of pneumatically propelled spheres. Certain crudes and some partially processed products are known to contain very low percentages of potentially corrosive agents, so where even minimal risk exists, it is now common practice to monitor internal conditions by the insertion of corrosion probes, and these may be augmented by the injection of chemical inhibitors into the flowing stream. Alternatively, in certain systems the pipe may be lined internally with an epoxide resin neutral to the fluids concerned.

The effects of corrosion on the exposed surface of buried metals, particularly those of ferrous origin, were known long before the use of steel for pipelines, or indeed the discovery of oil itself. The two practical methods of preventing or reducing the risk of corrosion to buried steel pipelines are (a) to prevent direct contact between the pipe and the electrolyte, i.e. the surrounding soil or water, (b) to reverse the electromotive force which causes loss of pipewall fabric.

The first may be achieved by providing an insulating barrier around the pipe. Even in the early days of cast and wrought-iron pipes attempts were made to reduce effects of corrosion by various methods of insulation. Earlier water engineers buried their mains in a chalk or sweet sand backfill, while gas engineers hand painted their services with tar and sand, or ran molten pitch into a disposable cardboard trough fitted around the pipe. Pipes protected in such fashion are known still to exist in the corrosive London clay, upwards of a century old and still be in excellent condition when exposed.

The modern pipeline wrapping differs little in principle from these earlier safeguards. Essential requirements are that it should be resistant to impact, soil stress, flow, water, electric current, bacteria, and marine organisms; possess good adhesion and flexibility; and be chemically stable. Molten coal-tar and bitumen, suitably reinforced with woven fabric, have been used for several decades for pipe wrapping, and they remain, perhaps, still the predominant selection today, although new materials are under constant investigation and development, or undergoing field trials. Earlier reinforcements included woven hessian, coir, and asbestos. The former pair tended to rot, the latter is mechanically weak and the normal present day counterpart is woven glass fibre. More recent techniques include extruded or sintered polyethylene, and thin-film epoxy coatings, in addition to the wide selection of self-adhesive tape wraps for hand or machine application.

A completely homogenous and continuous coating is essential, and every wrapping process is followed by a check with high-voltage detection equipment to ensure any voids are eliminated.

The electrical reversal of potentially corrosive currents is an application first extended to the protection of a pipeline in 1930, but the principle was known centuries ago when the electrochemical series of metals was determined. This series classifies each metal in a sequence of 'nobility'.

If any two metals are connected and placed in an electrolyte, a current will be generated and the less noble will corrode providing protection to the higher metal, i.e. the lower metal in the series becomes anodic and the higher, cathodic. Potassium, magnesium, aluminium and zinc are at the lower, or reactive end of the scale, while gold, platinum, and silver are at the higher, or passive end. Thus, if a mass of metal less noble than mild steel is connected to a buried pipe, it may be assumed that a current will flow from this mass into the pipe, which then becomes a cathode in the circuit, hence 'cathodic protection'.

Although the principle outlined still has many applications, particularly in the marine sphere, the sacrificial anode acting without artificially impressed motive power will provide only a very low protective current, of the order of a few milli-amperes, and it is now normal for the long cross-country pipelines, or indeed any other large steel structure, to be protected by means of an impressed current system, whereby low voltage D.C. from a transformer/rectifier is applied in conjunction with a carefully designed pattern of anodes known as a 'ground-bed'. Primary power supply may be taken from the local grid if available, or in remote areas wind-driven or solar generators, or batteries may be employed. Some very large lines which cross vast uninhabited tracts have dedicated generators with drivers powered by fuel from the pipeline itself.

Submarine pipelines may be cathodically protected in the same manner as buried land lines, sacrificial anodes being strapped at intervals around the pipe in the form of bracelets, usually of the same outer diameter as the concrete weight jacket to facilitate passage through the lay-barge tensioning equipment and to avoid vulnerable protrusions when on the sea bed. Alternatively, anodes may be held in suspension in a buoyant frame anchored adjacent to the pipe in similar fashion to a land-type ground-bed.

While either insulated coating or cathodic protection will provide individually a measure of protection against corrosion, it is normal practice to employ both in combination, thereby reducing any general corrosion of a buried pipe to the attainable minimum.

Third party damage

While the various mehods of prevention described for the preceding causes of spillage can be considered to be fairly consistently applied in all countries, there appears to be no standard approach to the problem of damage arising from work carried out by other parties in the vicinity of a buried pipeline, although there is similarity in several of the more obvious precautionary measures adopted.

Although this type of incident has been responsible for the major proportion of oil spillages, because of the excellent overall performance of oil pipelines, the actual number of accidents have still been small. Nevertheless, the nature of the damage inflicted has resulted in spillages of larger volume than is normally the case with other causes.

Some states have attempted to enact legislation or to produce definitive codes to help reduce such occurrences, but it has proved difficult to legislate without imposing burdensome penalties on pipeline operators as well as the offending other parties. Defined apportionment of liability has featured in some legislative proposals, but this aspect is probably covered in most instances by existing law or precedent, and is largely a matter for lawyers and insurance interests. The pressing concern in industry is to prevent damage occurring.

Usually, at least four separate interests are involved when considering potential damage to existing buried plant:
1. The owners/operators of the existing service;
2. The regulating or licensing authority to whom application must be made for the new works to proceed;
3. The promoters of the new works;
4. The contractor appointed to execute the new works.

Each of these has a very distinct role to fulfil towards the successful avoidance of damage as a result of the new works. The owners or operators of the existing service must be in a position to provide concise and accurate information to enable identification and location of their property. They should also, in their own self interests, be prepared to provide site attendance by informed and responsible operatives during the period of activity immediately adjacent to the plant concerned, cost reimbursement for such a service being an item for negotiation.

The regulating authority should provide the vital link between all parties, either by dissemination of information already collated from all service utilities and centrally recorded; or by providing a two-way communication and exchange of information between the various interests.

The promoter of the new works should make himself responsible for ensuring that enquiries are made of all owners of existing underground services, that their interests are safeguarded and their reasonable requirements for protection met. He should ensure also that his own intentions are clearly defined, and that sufficient notice is given. Too often promoters attempt to delegate these responsibilities to their contractors.

The contractor should cross-check, and be provided with tangible evidence, that the promoter has fulfilled all obligations with regard to exchange of information with other service companies, and he should be provided with details of all restrictions applicable to his operations, and precautions to be observed. Perhaps the most important contribution by the contractor is the instruction, supervision and control of his operatives. The man whose hand controls the damage-inflicting machinery must be as fully informed as all other parties.

Surface markers to indicate the presence of pipelines and cables are normally installed at boundaries, changes of direction, and other strategic locations. West Germany in particular has specific regulations for the positive location and identification of buried pipelines. However, it is not unknown for surface markers either to be incorrectly positioned, to be relocated wrongly during later surface works, or to be bodily removed by vandals. A very necessary adjunct therefore is production of accurate and detailed 'as built' drawings while construction is still in progress and all relevant depths and levels may be recorded precisely. Identifiable points on the pipe should be referenced back to fixed datum points to facilitate future location with accuracy, together with details of all other underground features and services exposed during the course of construction. In France a system has been introduced whereby drawings indicating existing pipes and cables form a necessary inclusion in the documentation required to perform works, and where the promoter is obliged to determine in advance who owns or operates each conduit.

Particular precautions are usually taken through developed areas and across footways, verges or roads, to provide additional warning of a pipeline's presence. These may include precast concrete slabs with indented lettering laid a few inches above and along the affected section of pipe, or alternatively a continuous plastic warning tape of colour appropriate to the service it protects, and with identifying labelling, may be used.

One novel safeguard for contractors' excavating equipment has been developed by the Canadian Gas Research Institute. A series of sensors have been fitted into the steel teeth of a back-hoe mechanical shovel, and these have proved capable of detecting a buried metal pipe towards which the bucket is directed and activating an alarm in the operator's cab. While apparently quite effective, it is preferable that such equipment be employed to supplement other precautionary measures rather than in isolation.

Information provided by regular line patrols has been of valuable assistance in preventing damage where un-notified or unauthorised works encroach onto a pipeline easement. These patrols are carried out on foot by line-walkers, and also by aerial observers, generally by helicopter in more populated areas and by light aircraft in remote regions.

As previously mentioned, regular inspections are also carried out on submarine pipelines; by divers in very shallow waters, and by manned or remotely controlled submersible craft in the deeper regions. The latter vessels are normally equipped with cameras and closed circuit television or recording videotape. Some evidence of damage to the concrete weight coating of surface laid lines by trawl boards has led to employment of a very high impact resistant coating, one type of which contains granulated Aberdeen granite. The precast concrete covers previously mentioned as a protection against current scour, are also a very effective safeguard against trawler damage, and have the ability to 'lift' the trawl gear safely over the pipe.

In some countries, the location of all underground services is now recorded by a statutory authority, and private property deeds may also be endorsed with relevant information, together with applicable conditions or restrictions. For instance, in the U.K. all pipelines are recorded in the local Land Registry and a solicitor's 'search' in relation to property transfer will reveal the presence of such plant. Some pipeline operators mail a prepaid reply form annually to each property owner or occupier along their pipeline routes to check for changes of tenancy. This practice serves also as a useful reminder to all concerned of the continued presence of the lines.

Many defensive schemes have been devised by individual undertakings or through the combined effort of groups of service operators. Several states in the USA operate a 'one-call' system by which means any intending excavator may obtain all information, or be provided with contacts, relevant to his proposed area of activity. One state has developed a system of graphic communication which operates in similar fashion to a telex network, except that transmissions are in diagramatic sketch form. Receipt of an enquiry with a rudimentary sketch of the affected area will result in responses from outstations detailing what plant is known to exist in that location.

In the U.K. Central and East Lancashire now have the services of JULIE (Joint Utilities Location Information for Excavators) to safeguard buried services in that region. A single call to this 'guardian angel' will provide all available information regarding pipes and cables in the subject area.

Many similar schemes are in operation or are under development, and emphasis is laid on the necessity for ensuring effective and constant communication between all involved parties and at all levels of responsibility, to ensure success.

CONCLUSION

From analysis of available data, four indisputable facts emerge:
1. Spillage of oil from pipelines is very infrequent.
2. Where spillage has occurred, damage has generally been localised and of a minor nature.
3. As far as Europe is concerned, over the twelve years during which pipeline spillages have been recorded, not a single case involving pollution of potable water sources has been reported.
4. The most significant influence on pipeline spillages worldwide, both as regards number of incidents, and volumes of oil lost, is that resulting from other works being carried out across or adjacent to buried lines, frequently referred to as third party damage.

In spite of an undeniably excellent record of environmental conservation, there is no indication of complacency in the oil pipelines industry, and efforts continue towards the development of further improvement to an already creditable performance.

Statistics collated over the past decade show that although there has been massive expansion of pipeline networks and a correspondingly vast increase in the quantities of oil transported, there has been no proportionate increase in the number of spillages reported each year. It is suggested that this is a reflection of the dedication and expertise of the pipeline engineers and operators, due in no small measure to a genuine concern for the environment.

REFERENCES

1. Stitching CONCAWE (The Oil Companies' International Study Group for Conservation of Clean Air and Water in Western Europe) *Annual Report on Spillages from Cross-country Oil Industry Pipelines in Western Europe.*
2. *Annual Report by Office of Pipeline Safety,* U.S. Department of Transportation.
3. *Annual Oil Pipeline Performance Review* by Canadian Petroleum Association.
4. Pipelines Act 1962.
5. Petroleum and Submarine Pipelines Act 1975.
6. British Standard Code of Practice. C.P.2010.
7. British Standard specifications. BS4515.
8. Institute of Petroleum:- Code of Safe Practice for Petroleum Pipelines.
9. (i) American National Standards Institute Publication No. B31.4 Liquid Petroleum Transportation Systems.
   (ii) American Petroleum Institute Code No. 1105 – Recommended Practice in Construction of Steel Pipelines.
   (iii) Codes No. AP1.5L, AP1.5LX, AP1.5LS – Specifications for production of line pipe.
   (iv) Code AP1.1104 – Field Welding of Pipelines.

9

Spills during transport by road and rail

Ian A. Wood BEng, MIME, FInst Pet
lately Esso Petroleum Co. Ltd

The transport of petroleum by road and rail in the United Kingdom is largely confined to refined petroleum products. There are a few movements of crude oil, but as a percentage of the total they are negligible.

Refined products range from light, volatile and highly inflammable substances such as 'petroleum spirit' or 'petrol' at one end of the scale, to the heavier, more viscous and less inflammable substances such as fuel oils, lubricating oils and bitumen at the other.

By far the greater part are carried in bulk — that is, in rail tank wagons or in road tank vehicles — but a small percentage, particularly of substances such as lubricating oils, are carried in drums and packages, and may be included with other general freight.

In this chapter reference is essentially to movements in bulk; the transport of liquefied petroleum gases is not included. This is because the transport of these gases is a specialised subject and any spillage or leakage of them creates great hazards, which cannot be adequately dealt with in a general review. The need for some regulation has been highlighted by the tragic disaster at a Spanish camping site in the summer of 1978.

Road transport is the means by which the greater percentage of petroleum products are finally delivered to the customer. Only large consumers having major demands are normally supplied by pipeline, water or rail. On the other hand, it is relatively unusual for road transport to be the only mode involved in moving products from refinery to consumer. It is more usual for a secondary tier of distribution — local depots or installations — to intervene as the supply source for road deliveries, and these depots and installations may in turn be supplied by rail, water or pipeline from refineries.

The method of delivery used is generally a reflection of the nature of the requirement. The average rail tank wagon movement is over a distance of around eighty miles, whereas the average road tank truck delivery distance is around thirty miles. The modern rail tank wagon is either a two axle wagon of about 45 tons gross weight, carrying perhaps 8,500 gallons (38,000 litres) of petroleum, or a bogie tank wagon of 90 to 100 tons gross weight, carrying perhaps 18,000 gallons (82,000 litres). Such wagons are, nowadays, predominantly operated as complete trains of perhaps 500 to 1,000 tons or more, and individual wagon movements only comprise a small percentage of the total. Furthermore, rail tank wagons almost invariably have single compartment tanks — that is, the tank is not subdivided — so the minimum delivery economically possible is around 8,500 gallons (38,000 litres).

Road vehicles on the other hand are, under present legislation,[1] restricted to 6,600 gallons (30,000 litres) capacity and, depending on the nature of the product to be carried, have frequently to be subdivided into compartments not exceeding 1,100 gallons (5,000 litres) capacity. It is thus apparent that road tank vehicles have greater inherent flexibility and are able to cater for smaller deliveries and mixed loads.

REGULATORY CONTROLS

The conveyance of petroleum products has been subject to regulatory control for a century or more. In the case of transport by rail, the requirements are those of the British Railways Board. In the case of road transport, current legislation stems from the 1928 Act.[1]

Basically, the degree of control varies according to the hazard assumed to be associated with the product. This 'hazard rating' is usually expressed in terms of the 'flash point' of the product, (see chapter 10) this being the lowest temperature at which the application of a small flame causes the vapour above the product to ignite, when the product is heated under prescribed conditions. In the United Kingdom, legislation defines any petroleum which gives off an inflammable vapour at a temperature of less than 73°F (approx. 23°C), when tested in the manner prescribed in the Act,[1] as petroleum spirit, and particular requirements apply to the conveyance of such petroleum spirit by road or by rail, as previously indicated.

Less stringent requirements apply to the conveyance by rail of petroleum products having higher flash points, between 23°C and 61°C. Conveyance by rail of petroleum having a flash point above 61°C is not considered hazardous.

There are, at present, no corresponding regulations for conveyance by road, within the United Kingdom, of petroleum products having a flash point above 73°F, but for international traffic, vehicles travelling to Europe are required to comply with the requirements of 'ADR' — the European Agreement Concerning the International Carriage of Dangerous Goods by Road — which prescribes various requirements for petroleum having flash points of below 21°C; from 21°C to 55°C inclusive, and from above 55°C to 100°C, respectively.

Regulatory and quasi-regulatory controls also exist concerning the labelling and marking of rail wagons and road vehicles, for the purpose of indicating the nature of the contents. In the case of road vehicles, the Act[1] itself requires that the words 'petroleum spirit' and 'highly inflammable' be displayed in conspicuous characters on the tank. Regulations made in 1971[2] require that a red diamond shaped notice containing a black flame symbol and the word 'inflammable' be displayed at the front and rear of a vehicle conveying petroleum spirit.

In the case of railway wagons, the British Railways Board requires that a highly inflammable wagon label be affixed to any rail wagon conveying petroleum spirit and an inflammable wagon label to a rail wagon conveying petroleum having a flash point between 23°C and 61°C.

More recently, the Hazardous Substances (Labelling of Road Tankers) Regulations 1978 have come into operation. These supersede previous marking requirements for road tankers. The principles of marking involved have also been adopted by the British Railways Board for use on rail tank wagons. The regulations require the use of a hazard warning panel at the rear and both sides of a road vehicle tank. This panel identifies the material conveyed by means of a 'substance identification number', and incorporates an 'emergency action code', indicating action to be taken by the emergency services. It also includes an appropriate hazard warning 'diamond' sign and a space for a telephone number or other approved text indicating how specialist advice concerning the substance carried may be obtained. These regulations also apply to multi-loads of substances presenting differing hazards. However, in the case of petroleum products there are special provisions to allow for the conveyance of certain compatible substances without the complication of multiple labelling.

VEHICLE DESIGN FOR SPILLAGE PREVENTION DURING TRANSIT

There appear to be no published statistics on spillages of petroleum products during conveyance. However, such information as is available indicates these to be at a very low level. They can be divided into two categories: those which occur whilst the vehicle is in transit, and those which occur whilst loading or unloading. It should be emphasised that the scope of the Conveyance by Road Regulations[3] extends to the loading and unloading of the vehicle concerned. From the available information, it seems likely that the incidence of spillage for rail tank wagons in transit, due to accident, etc is as low as 0.0001 per cent, whilst for road tank vehicles, the incidence is somewhat higher at around 0.0002 per cent. It would appear that the incidence whilst loading and unloading — particularly the latter — is several times as great, but on the other hand such spillage usually occurs in premises where some sort of catchment exists and recovery of the spillage is facilitated.

These low spillage levels are due to the fact that both rail and road tank vehicles are constructed in such a way as to minimise the possibility of loss in the event of accident. Rail tank wagons are designed to comply with the railways company's standards. They are substantially constructed of steel and although not necessarily designed to an acknowledged pressure vessel code are, nevertheless, effectively pressure vessels and are pressure tested. They are arranged for top loading through a manhole or manholes, which are of substantial construction with secure closures to withstand pressure. Where petroleum spirit is to be carried, venting is catered for by pressure-vacuum valves which will retain the contents of the tank in the event of overturning. Unloading is usually through bottom outlets, with valves to which a hose or similar connection can be attached, on either side of the wagon. These external valves and their associated piping draw product through internal footvalves, and the pipework external to the tank is provided with a shear section adjacent to the footvalve which will break off under severe impact or strain, leaving the internal footvalve intact. Thus a rail tank wagon should be capable of retaining its contents in all but the most severe accidents.

Road vehicle tanks vary in design according to the nature of the product to be transported. Where petroleum spirit is to be carried, then the vehicle must conform to the requirements of the national regulations: in the UK these are the Petroleum Spirit (Conveyance by Road) Regulations.[3] These regulations call, *inter alia*, for individual compartments of tanks not to exceed 1,100 gallons (5,000 litres) capacity, and for the total capacity of the complete vehicle not to exceed 6,600 gallons (30,000 litres). The design of the vehicle must comply with the requirements of the regulations and, if over 1,500 gallons (6,800 litres) capacity, must be approved by a government inspector. This approval extends to the materials which may be used for the tank; the height of the centre of gravity, which is limited in order to promote stability on the road; the protection of tank top fittings by a coaming or by recessing them within the tank shell, to avoid damage in the event of overturning and, as in railcars, a requirement for internal footvalves, protected by an external shear section so that the internal valve may remain intact in the event of accident, and the use of venting valves which will retain the contents of the tank if the vehicle overturns. The regulations themselves furthermore require that the fill openings and draw-off openings be securely closed and locked, and any openings for dipping be securely closed, whenever the vehicle is in motion.

It will be seen that these requirements result in a vehicle which should not give rise to spillage in transit except under exceptional circumstances of severe impact. No similar United Kingdom legislation extends to vehicles conveying petroleum having higher flash points than 73°F, although the ADR requirements apply to vehicles bound for European destinations. In practice however, for convenience, other higher flash point products such as kerosine, gas oil, and particularly automotive diesel oil are frequently carried in petroleum spirit vehicles. Nevertheless these latter substances may be carried in single compartment tanks of a size limited only by the permissible vehicle weight, and 'black' products, such as fuel oils, are normally carried in such tanks. These are sometimes constructed as pressure vessels, using air pressure to discharge them, and in any event are robustly built and are no more likely than petroleum spirit vehicles to give rise to leakage in the event of accident.

FACILITIES FOR PREVENTION OF SPILLAGE WHILST LOADING

As has been indicated, incidence of spillage whilst in transit is small. The greater proportion of spillage occurs whilst unloading, and to a lesser extent whilst loading, tank wagons and road vehicles. Although equipment failure may contribute to such incidents, human error is much the most significant cause, and it is important that all personnel be well trained, competent and aware of their responsibilities. Whilst this is generally true of the personnel carrying out loading operations at an oil company's premises, the same may not be true at the customer's premises, where receiving oil deliveries may only be one of many functions of the individual concerned.

Top loading of rail tank wagons is normal. Although there are many multi-point facilities in use, i.e., facilities with one fill pipe per wagon (or two in the case of bogie wagons), such facilities are in general being displaced by single point loading facilities. The multi-point facility, by means of a series of valve manifolds and fill pipes each spaced one wagon length from the next, makes it possible to load any available product into any wagon without moving the train. This system permits the loading of a number of wagons at the same time, but in terms of the individual wagon, the loading rate is relatively slow, thus giving the operator time to control the topping-up of each wagon. It is, however, almost entirely dependent on the competence of the operator so far as the avoidance of spillage is concerned, and since the operation is spread along the whole length of the train, drainage to cater for spillage is both expensive and of varied effectiveness.

Single point loading facilities, on the other hand, concentrate a number of fill pipes or lances in one location and the train is drawn past this location, wagon by wagon, by a cable operated 'wagon mover,' the correct grade of product being loaded through the appropriate fill pipe. This system has the advantage that, although individual wagon loading rates are high, accurate metering is facilitated, and safety devices to prevent overfilling and accidental wagon movement can be built into the system. Furthermore, the relatively small area where filling is concentrated lends itself to effective drainage and interception of any spillage which may occur *(see plate 19)*.

Whatever system of loading is used, it is necessary to prevent the accidental movement of wagons by a locomotive or other external agency whilst filling is in progress. No locomotive should remain connected to wagons being loaded and a notice, flag, lamp or preferably a gate or points which can be locked to prevent access to the siding are desirable.

Loading of road vehicles, on the other hand, is normally undertaken by the driver of the vehicle. Top filling is the general practice. The vehicle is positioned alongside a loading platform on which are mounted a series of loading arms for the various grades

Plate 19: *Rail loading area at BP's Kent refinery.*

Plate 20: *Loading a road tanker.*

of product. The appropriate loading arm or arms are positioned by the driver with the drop pipe through the filling opening. (*see plate 20*). Where loading rates are high, the product is normally metered into the vehicle, although in small plants the vehicle may still be loaded against the dip stick in the vehicle compartment. Heavier products, especially if handled when heated, may be loaded by weight.

It is normal practice for the driver to set up the quantity required on a pre-set meter or meters, which will shut off flow when the correct quantity has been delivered. The control valves and pumps for such meters are normally linked to devices in the vicinity of the loading position which permit of immediate emergency stops by pulling a trip wire or pressing a button and which, in some installations, may also be actuated by high level cut-offs in the vehicle tanks in the event that overfilling occurs. The meters may also be connected to a remote read-out in a control centre and, not infrequently, may also be electrically linked to security devices to prevent operation by unauthorised personnel.

Despite these precautions, spillage can still occur and the vehicle loading areas at loading facilities are normally surfaced with an impervious material, such as concrete, and the surfacing is graded to lead any spillage away from the vehicles to drainage passing through an interceptor, in which spillage can be retained and recovered.

FACILITIES FOR PREVENTION OF SPILLAGE WHILST UNLOADING

Unloading of rail vehicles is normally the responsibility only of the personnel at the point of discharge, who will generally have control over the whole of the facility from the rail siding to the storage tank and will be able to check the grades and quantities to be received into storage and the available capacity for each grade. Except in the rare instances where gravity discharge is possible, the discharge will be effected by fixed pumps on the premises where the unloading is to take place.

Unloading facilities for trainload movements normally involve multi-point discharge, with a hose or other flexible connection at each wagon position, connected via a manifold into discharge pipes for each grade. Reverse flow of product through the hoses is prevented by non-return valves.

It is important that, the wagons having been positioned in the siding, the engine be detached and, as when loading, some positive means — preferably a lockable gate or points — be used to prevent its return until unloading is complete; all hoses have been disconnected, and the wagons made ready for removal.

Care must be taken when disconnecting hoses after unloading to avoid any slight spillage of product. Facilities for 'slopping' any small quantities so released and drainage to catch any slight spillage and lead it to an interceptor are desirable, since tank wagon unloading hoses are not normally equipped with self-sealing couplings.

Responsibilities for unloading road vehicles are not generally so simple as in the case of rail tank wagons. They may well be divided. The procedures to be followed in the case of vehicles carrying petroleum spirit are laid down in regulations.[3] Basically, they require that the vehicle be constantly attended by the driver or other competent person; that the vehicle engine shall not be run whilst unloading; that the licensee of the premises must ensure that a competent person other than the person in attendance on the vehicle is present, who must ensure that capacity exists in the storage for the grade to be delivered, and that the discharge hoses are securely connected to the correct storage tank. A certificate is prescribed, to be signed by the recipient before discharge may be commenced.

Where the product to be unloaded is other than petroleum spirit, no such regulations exist and the responsibility may vary from location to location. Sometimes the driver of the vehicle may check and ascertain the capacity of the storage, connect up the hose, and effect the delivery himself. This is normal, for example, in the case of deliveries of domestic heating oil. *(See plate 21)* However, where there is more than one grade of product to be delivered or, for example, the storage is not adjacent to the vehicle, it is normal for the recipient to have a competent person in charge of the storage, to ensure safe delivery.

The facilities existing at the unloading point of a road vehicle may be very varied. At a petrol filling station or garage, tankage will normally be below ground, and discharge will be effected by gravity. In industrial premises having tankage of small capacity, a similar situation may prevail. However, if storage is of larger capacity, tankage may be above ground, in which case, if petroleum spirit is to be delivered, a suitable independent pump will be necessary to unload the vehicle. However, for other products such as kerosine, heating oil, fuel oil, etc. it is frequently possible for the vehicle itself to effect discharge by its own engine-driven pump or, in the case of a vehicle having a tank designed as a pressure

192 . *Prevention of Oil Pollution*

vessel, compressed air from a compressor on the vehicle may be used to expel the contents.

OPERATING PRECAUTIONS

It has been indicated that spillages occur either in transit or whilst loading or unloading. Those which occur in transit are usually associated with an accident to the vehicle. Those which occur whilst loading and unloading are, on the other hand, for the most part avoidable if personnel are properly trained and due care is taken. Failure to observe the following precautions accounts for a substantial proportion of spillage during loading.

a. Safety precautions for vehicle loading
    i Loading arms, meters, control valves, etc. should be marked with the grade of product which is delivered through them.

Plate 21: *Road tanker which delivers and meters domestic fuel oil.*

ii If there is any system of remote control of loading then the vehicle loading position should be identified by a number.

iii There should be adequate emergency stop controls available to loading personnel, both at the point of loading and, in case of fire, clear of it.

iv Any compartmented vehicle tanks should have compartment numbers marked adjacent to the filling opening for the compartment.

v Identical numbers should be marked near the outlet valves for each compartment, and on any bottom operated footvalve controls for internal footvalves.

vi The capacity of each tank or compartment should be marked near the filling opening.

vii Road vehicle dipsticks should be clearly marked with the vehicle number, compartment number and unit of measurement. (Rail tank wagons are measured with an ullage stick, which is a measure of the distance from the tank top to the surface of the liquid and hence individually calibrated ones are not required for each tank).

b. Reasons why spillage occurs when loading

i Failure to check that footvalves and outlet valves are closed.

ii Lack of awareness of tank or compartment capacity.

iii Failure to check whether tank or compartment is empty.

iv Setting wrong quantity on loading meter.

v Putting loading arm in wrong compartment.

Similar precautions apply when vehicles are to be unloaded. The precautions noted here apply particularly under the conditions that would exist at an industrial location where receipt of petroleum is ancillary to the main function, and where the facilities are of a fairly simple nature. Similar considerations apply at major petroleum storage installations, but the facilities might be more sophisticated.

c. Safety precautions for vehicle unloading

i Every tank or tank compartment used to store product should be clearly numbered, the number being in a position where it can best be seen by the person responsible for making the delivery to the tank.

ii All tanks should have a dipstick or tank gauge, showing the tank capacity, and clearly marked with the number of the tank with which it corresponds.

iii If there are individual offset fill connections, not obviously related to the tanks, and particularly if the tanks are underground, the fill point should be clearly marked with the number of the tank with which it corresponds.

iv Fill connections should preferably be unique to a single tank, or to two or more tanks storing the same grade of petroleum, and should be marked to show the grade concerned. Where such segregation is not possible, additional instructions, notices, diagrams, etc. may be necessary.

v It may, in certain cases, be desirable to equip tanks with audible high level alarms or even with automatic cut-offs.

vi Hoses should first be connected to the filling point to which delivery is to be effected, then to the outlet connection of the delivery vehicle. Valves between the hose and the receiving tank should then be opened and lastly the outlet valve of the road or rail vehicle.

vii If unloading is to be effected by means of a pump, pumping may now commence. A competent person or persons, including the driver of any road tank vehicle involved, should keep a constant watch during the whole of the time of delivery and should be alert to stop product flow in case of emergency.

viii When unloading is complete, 'dry' hoses — i.e., those which are normally empty during transit — should be drained or 'blown through,' if possible, into the receiving tankage; any valve to the receiving tank or pipeline closed; the outlet valves of the road or rail vehicle closed and the hoses disconnected, care being taken to catch any product remaining in the hose and dispose of it without causing a spillage.

ix The area where unloading of road vehicles is to take place should preferably be surfaced with an impervious material and be graded and drained to a suitable interceptor so that any spillage can be retained and recovered.

d. Reasons why spillage occurs when unloading
  i   Incorrect or unnumbered dipsticks or gauges causing errors in measurement.
  ii Connection to wrong storage tanks caused by inadequate identification of fill connections.
  iii Connection to wrong vehicle compartment, caused by carelessness or inadequate marking on vehicle outlet valves.
  iv Bursting of hoses, due to failure to open all valves to storage tanks before starting vehicle pumps, or use of hoses which have become defective, possibly due to lack of care in handling or improper stowage, and which have not been properly inspected and tested.
  v   Overfilling storage, due to lack of care of personnel involved, when only a partial discharge of the delivery vehicle is to be effected.

ACTION WHEN SPILLAGE HAS OCCURRED

Certain basic action is necessary, largely common to road and rail, when an accident or spillage occurs in transit. The operator of the equipment — the oil company or vehicle owner concerned — or the railway authority in the case of rail transport, should be notified immediately and, if an emergency situation exists, appropriate action should be taken. In the United Kingdom the police have an overall responsibility for general control of any emergency and will also notify other emergency services, such as the fire service, local and water authorities, as necessary.

In the event of a railway accident, the railway authority is responsible for providing access to the site; recovering of rolling stock; repairs to track, but will normally call on the services of the oil company operating the tank wagon to assist in handling spillage, evacuation of contents, etc.

Major oil companies, apart from having direct responsibility for their own road vehicles involved in accidents, have personnel and equipment available to deal with emergencies, and can offer technical advice and assistance in containment and recovery of product; cleaning up spillages and general control of pollution.

Water authorities are concerned with any spillage which may enter waterways or lakes, or which is likely to contaminate ground water resources or enter sewers or drains.

When a spillage occurs not in transit but in premises where the vehicle is being loaded or unloaded, the situation may be rather different. Particularly if the premises are licensed to store petroleum spirit, the conditions of licence may specify additional requirements which have to be complied with, such as notifying the licensing authority.

Where loss of life or personal injury is occasioned by explosion or fire involving a vehicle conveying petroleum spirit, the owner of the vehicle in which the petroleum spirit is being conveyed, or into or from which it is being loaded or unloaded when the fire or explosion occurs, must immediately notify the Secretary of State concerned. In practice, this means that the Health and Safety Executive must be notified.

The principles of dealing with spillages from vehicles are basically similar to those for any other type of spillage. Where possible, the source of the leak should be sealed off, by closing any open valve or by any other available means. Every attempt should be made to contain the spillage, for example by blocking off drains; collecting the leakage in containers; digging a collection pit, etc. Where highly inflammable product is involved, any adjacent open fill covers, valves, etc. should be closed; the area cleared of bystanders and any possible source of ignition eliminated. If it is necessary to remove vehicles from the area, they should be pushed or towed. The engines should not be started. Once the spillage has been contained, steps should be taken to remove the spilled product and to evacuate the contents of leaking vehicles as quickly as possible, but the method employed should not be such as to create any possible source of ignition if the spillage is of a highly inflammable nature.

Small amounts of spillage — either small spills or amounts remaining after disposal of larger spillages — may be mopped up by covering with sand; one of the various commercially available absorbent materials, or even straw in the case of the more viscous products. Such absorbent material may be removed to a safe area and disposed of as circumstances permit, for example by incineration or dumping at an approved location.

REFERENCES
1 Petroleum (Consolidation) Act 1928. Under this Act Regulations for road transport have been made.
2 Inflammable Substances (Conveyance by Road) (Labelling) Regulations 1971.
3 Petroleum Spirit (Conveyance by Road) Regulations 1957 as amended by the Petroleum Spirit (Conveyance by Road) Regulations 1966.

10

Domestic and industrial storage

H. Jagger, B.Sc, FInstPet
lately Esso Petroleum Co Ltd

DOMESTIC AND INDUSTRIAL STORAGE

Earlier chapters have indicated that petroleum products are extensively used as vehicle fuels, for heating purposes in both domestic and commercial premises, as lubricants and as feedstocks to chemical and other processes. In considering domestic and industrial storage we are concerned with the full range of products from liquefied petroleum gases (LPG) which is used domestically as well as for heating in horticulture and chick rearing and in a range of industrial applications, through naphtha or light distillate feedstock (LDF) which is used as chemical feedstock, petrol, paraffin (kerosine), aviation fuels, gas oils and diesels to residual fuels which are used extensively for power generation, as industrial fuels and for heating larger domestic, commercial and other premises and including crude oil stored in bulk at refineries. With such a large range of products to store, it is clearly necessary to have some knowledge of the properties of each class of product and of the type of storage most suitable to each class, if spillage and consequent damage to the environment — to say nothing of the loss of costly materials — is to be avoided or at least minimised.

CLASSIFICATION OF PETROLEUM PRODUCTS

Petroleum, or crude oil, is a complex mixture of compounds of carbon and hydrogen with traces of impurities such as sodium, vanadium, sulphur and other elements. The lightest gaseous compounds, methane and ethane, are removed as natural gas at the oil well before the oil is transported to refineries to be further divided and processed to yield both the heavier gases, propane and butane, which are liquefied under pressure as LPG, and the full range of liquid products. Because these liquid products are themselves mixtures of a number of different hydrocarbon compounds, they evaporate, and on further heating they eventually boil, over a range of temperatures rather than boil at a specific temperature as does a single pure compound such as water. It is the vapour which is given off by the hydrocarbon liquid, and not the liquid itself which burns on ignition, so that the lighter and more volatile compounds which give off vapours readily at normal storage temperatures are readily ignited while heavier products require heating to produce sufficient vapour to be capable of ignition. In any case, the vapours will only ignite when mixed with air in certain proportions. If the percentage of vapour in the mixture is less than one per cent the mixture is too lean to burn and is said to be below the lower flammability limit. When the mixture contains more than ten per cent of hydrocarbon vapour (for propane and heavier hydrocarbons) the mixture is too rich to burn and is said to be above the upper flammability limit. The temperature to which a substance must be heated to give off sufficient vapour to form a mixture with air that can be momentarily ignited, is its flash point. The fire point is the slightly higher temperature at which the substance gives off sufficient vapour to burn continuously after the test flame is removed. Volatile products such as LPG, naphtha and petrol have flash points well below freezing point. For petrol it is -45°C. Paraffin has a flash point above 21°C while gas oils, diesel fuels and residual fuels have flash points above 55°C.

The following classification is used in the petroleum industry and by regulatory bodies in considering the fire risks and regulations appropriate to the storage of a particular petroleum product:

Class 0 Liquefied petroleum gases
Class I Products which have a flashpoint below 21°C
Class II Products with a flashpoint from 21°C up to 55°C
Class III Products which have a flashpoint from 55°C up to 100°C
Unclassified Products with a flashpoint above 100°C

In cases where products are stored or handled at temperatures above their flashpoint they should be treated as Class I products even though they normally fall into another classification.

STORAGE HAZARDS

The major hazards encountered in the storage of Class 0 and Class I products are the fire and explosion risks resulting from the ease with which these products evaporate and their low flashpoints. Because the Class 0 products (LPGs) are gases under normal conditions of pressure and temperature, they can be stored only under pressure at ambient temperature or at refrigeration temperatures under near ambient pressures. Except for very large storage installations, they are normally stored in either cylindrical or spherical containers at ambient temperatures. Special care is required in the initial filling of such containers to remove air and so avoid the possibility of forming an explosive gas/air mixture in the container. Care must also be exercised to avoid overfilling the pressure container and so leaving insufficient space for thermal expansion of the liquefied product. For the portable containers normally used for both the transportation and storage of these products this is achieved by filling to a specified maximum weight for each container and type of product, the weight being calculated

Plate 22: *Esso terminal at West London. Each of the six tanks in the foreground has a small bund, with common larger one around them all.*

to ensure sufficient vapour space for safety under the worst expected conditions. Any leakage from an LPG container will be of vapour if the leak is above the liquid level, or of liquid which will evaporate quickly at first and then more slowly as the liquid temperature is reduced by the evaporation, if the leak is below the liquid level. In either event a flammable and possibly explosive mixture of gas and air will form unless there is sufficient air movement to ensure complete dispersion. Continued leakage may result in an accumulation of gas at a low point near the leak, the centre of the accumulation being of gas with little air and therefore too rich to ignite but surrounded by a mixture of gas and air within the explosive range. To minimise the chances of such a situation arising Class 0 products should be stored in well ventillated locations chosen to avoid the possibility of escaping gas drifting towards a source of ignition. The fire and explosion risks associated with the storage of Class I products are the main considerations of the Petroleum Spirit Storage Regulations in the United Kingdom.[1] They include the requirement that 'petroleum spirit,' which means liquid petroleum products with a flashpoint below 73°F, shall not be stored except in facilities licensed by the appropriate local authority. Local authorities are guided in their considerations of licence applications by the Home Office Model Code which sets out the detailed requirements for the construction and operation of petrol storage installations from major oil company terminals and storage depots to retail stations.[2] While the requirements for small user installations are less stringent than for large storage installations in view of the smaller quantities involved, the same considerations are involved to minimise the risks of fire resulting from spillage or from carelessly located tank vents. These requirements apply to the storage of Class I products in cans, drums or other containers, other than the fuel tanks of vehicles, as well as to storage in conventional tanks. For Class II and heavier products the fire risks are considerably less and environmental risks constitute the more important considerations.

TYPES OF TANKS

LPGs are most frequently used for small consumption units such as domestic cookers, as fuel for forklift trucks used in confined spaces and for small industrial uses. The cylinders used to transport the fuel also serve as storage containers from which the fuel is dispensed through a pressure reducing valve and control cock suited to the appliance. For larger consumption installations, where delivery may be by bulk road or rail vehicle, specially designed pressure storage vessels must be used to withstand the vapour pressure of the fuel at the highest storage temperature likely to be experienced. The alternative of storing the LPG at atmospheric pressure and employing refrigeration to reduce the vapour pressure is used only in situations where large quantities are stored for long periods and used irregularly — a situation unlikely to apply to user installations. In either event, vessel design should be to a suitable design code and reference should also be made to national codes.[3,4]

For petroleum liquids, three types of tank are commonly used. Welded rectangular steel tanks with capacities up to 3,500 litres (750 gallons) are used for domestic and small industrial or commercial heating installations. Galvanised tanks should not be used. The size of tank should be such as to allow deliveries to be made in economic parcel sizes and allow a margin of several days consumption. Larger industrial installations whether for gas oil, diesel fuel or fuel oils will have tanks between 3,500 litres and 50,000 litres (12,000 gallons) and these will normally be horizontal cylindrical tanks of welded steel construction. Appropriate design standards should be observed.[5,6]

Where quantities larger than 50,000 litres are stored, vertical cylindrical tanks are used. These are almost invariably of welded steel construction although a few rivetted tanks are still in use and sectional bolted tanks have been installed at sites where special circumstances such as urgency of requirement or a lack of skilled welders ruled out conventional tanks. National Standards covering the requirements for vertical steel tanks for petroleum product storage frequently provide for three categories:[7]

1. Unpressurised tanks suitable for the storage of Class III products.
2. Pressurised tanks to be fitted with pressure and vacuum relief valves and suitable for the storage of Class I and Class II products.
3. Floating roof tanks in which a pontoon or other type of roof floats on the stored liquid thus reducing the loss which results from the breathing which takes place when a fixed roof tank is filled or emptied and as a result of temperature changes.

Floating roof tanks are preferred for the storage of Class I products and for the storage of Class II products stored above their flashpoint, since by reducing vapour loss they both save costly product and reduce fire risks.

Tank fittings

All tanks should be fitted with connections for filling, for product draw-off, gauging, sampling, venting and draining, although on smaller tanks some of the connections may serve more than one of these requirements. Thus a single screw capped dip hatch on a domestic fuel tank will also serve as a sample hatch and as a vent during filling. The product filling line on small tanks should enter the top of the tank and discharge near the bottom of the tank and be provided with an anti-syphon hole to prevent back-flow of product.

All connections should be in accordance with the design standard applicable to the tank and should be of steel. Those connections which are below the liquid level when the tank is full should be fitted with steel valves as close as possible to the tank shell. Filling connections for tanks used for storing Class I and Class II products should terminate near the bottom of the tank to reduce splashing and the possible generation of static electricity. The product draw-off line should be raised sufficiently from the bottom of the tank to avoid drawing water, sludge or sediment into the system being fed from the tank. The drain connection should be from the lowest point of the tank to facilitate complete removal of water and to permit complete removal of product prior to cleaning or change of product to be stored.

Tank vents should be sized to ensure that the design conditions of pressure and vacuum are not exceeded during tank filling and emptying. On small tanks and on large tanks of the non-pressure type, open vents protected with a coarse wire mesh may be used, although in the case of petrol tanks these should extend 3.75 metres above ground level and discharge upwards not less than 1.5 metres from any window or other opening in adjacent buildings or from the boundary of the property in which the tank is located. For larger tanks containing Class I or Class II products, and therefore designed to withstand some pressure and vacuum, pressure and vacuum valves are fitted and set to release before design conditions are reached. These valves are fitted directly onto the tank with no intermediate shut-off valve and should be designed to take care of maximum filling and emptying rates together with an allowance for breathing due to temperature changes. Provision should also be made for emergency venting in case a tank is subjected to excessive temperatures due to fire. In the case of vertical tanks this emergency venting is provided by the floating roof where this type of tank is used. On fixed roof tanks a specially designed weak welded seam between the tank wall and tank roof ensures that the roof will blow off before the circumferential wall is damaged with a consequent loss of product.

Gauging connections to tanks vary with the size and type of tank and with the type of gauging to be used. For small domestic or other tanks in paraffin or gas oil service, a simple glass or clear plastic gauge tube connected to a valve near the base of the tank, secured to the side of the tank and extending above the highest liquid level, provides a visible indication of the oil level in the tank. For larger tanks, a pressure sensing element in the bottom of the tank may be connected to a suitably calibrated gauge, or a float on the liquid surface connected to a level indicator, will each require appropriate openings in the top of the tank. In either case a capped or hinged dip hatch should be provided to permit check dips to be taken using a dip stick or dip tape. For underground petrol tanks, as for small above ground tanks, the capped fill pipe also serves as the dip hatch. In the case of petrol tanks particularly, a reinforcing plate should be welded onto the inside of the tank immediately below this dip pipe to prevent damage to the plate and removal of any protective coating through repeated dipping.

For larger vertical tanks one or more dip hatches are provided on the tank roof near the outer wall, that near the access ladder or stairway being used for routine gauging and sampling while others should be available for checking purposes — including confirmation that uneven settlement has not occured. On floating roof tanks provision should be made for dipping through a vertical tube installed near the platform on top of the tank wall by the access stairway and thus avoiding the necessity of descending onto the floating roof. Automatic liquid level and temperature measuring instruments for vertical storage are desirable and should be installed according to the recommendations in the appropriate National Standard.[8]

Tank location, foundations and pipework

Storage of LPG does not constitute a pollution hazard in itself but because of the fire risks involved LPG should not be stored in the same bunded area as other products. LPG tanks should be located in well ventilated areas away from possible sources of ignition and from boundary fences.

The location of tanks for liquid products is important if the effects of spillage and leaks are to be minimised. The storage of products, other than Class I, is not normally subject to legislative control. In the United Kingdom only planning permission is required, except in the area of Greater London which was originally the London County Council area. Tanks for Class II and Class III products should preferably be located outdoors but small tanks may be located in buildings in fire resistant chambers vented to the open air. Such tanks should be supported on well constructed brick or concrete cradles designed to minimise the accumulation of water in contact with the tank shell and should be secured to prevent movement due to high winds or flooding. Easy access should be provided to all valves and for external inspection and painting. Outdoor tanks should be located so as to minimise the risk of damage from traffic or from other causes, to minimise the length of pipework required to convey the fuel to the boiler or other user equipment and to provide easy access for the fuel delivery vehicle. Where possible the fuel tank should be visible from the road vehicle when fuel is being delivered. In cases where a remote filling point is necessary the line should be valved close to the hose connection and be provided with a non-ferrous screw cap secured to the line by a chain. For many housing estates a centrally located storage tank is used to provide a metered fuel supply through a pipework distribution system. The special considerations involved in designing and operating such systems to minimise the risks of water pollution are the subject of the Institute of Petroleum's Code of Safe Practice, Part 10.[10] It contains details on tank location and installation as well as guidance on the special problems encountered with the extensive pipework associated with such installations.

The requirements of smaller commercial and industrial users will generally be met by the use of one or more horizontal cylindrical tanks. As with the small domestic tanks, these should be located with due consideration of the access and shortest possible piping requirements and should be sited away from streams or open drains. They should preferably be on a concreted area with well designed concrete cradles. Tanks should be clearly marked as to the type of fuel contained and delivery hose connections, when these are not on the tank itself, should be clearly marked with both the grade of fuel and the tank number.

Where larger quantities of fuel are to be stored and vertical cylindrical tanks are installed, carefully constructed tank foundations are required which elevate the tank slightly above ground level, avoid any uneven settlement and provide drainage and corrosion protection for the bottom plates of the tank. The foundation should be surfaced with bitumen-sand or similar material so that any oil leak appears at the edge of the foundation rather than soaking into it. Appropriate national standards should be consulted.[7] Where poor soil conditions are encountered consideration should be given to the use of concrete support rings or in extreme conditions to piled foundations to ensure that uneven tank settlement does not result in tank shell distortion or damage to connecting pipework.

The storage of petrol and other Class I products must comply with the local regulatory requirements which in the United Kingdom are set by the licensing authority to comply with the Petroleum (Consolidation) Act. These normally require that, for service stations and privately operated petrol pumps, the storage tank is underground, the tank being previously protected to minimise corrosion and surrounded by dry sand or other inert material to further reduce corrosion risks. The ground surface above buried tanks should be suitably reinforced to protect the tank and fittings and to allow traffic movement where necessary. *(See figure 23)* Since buried tanks cannot be inspected for corrosion or leaks, periodic pressure tests may be required on the tank

Figure 23: *Underground petrol storage tank.*

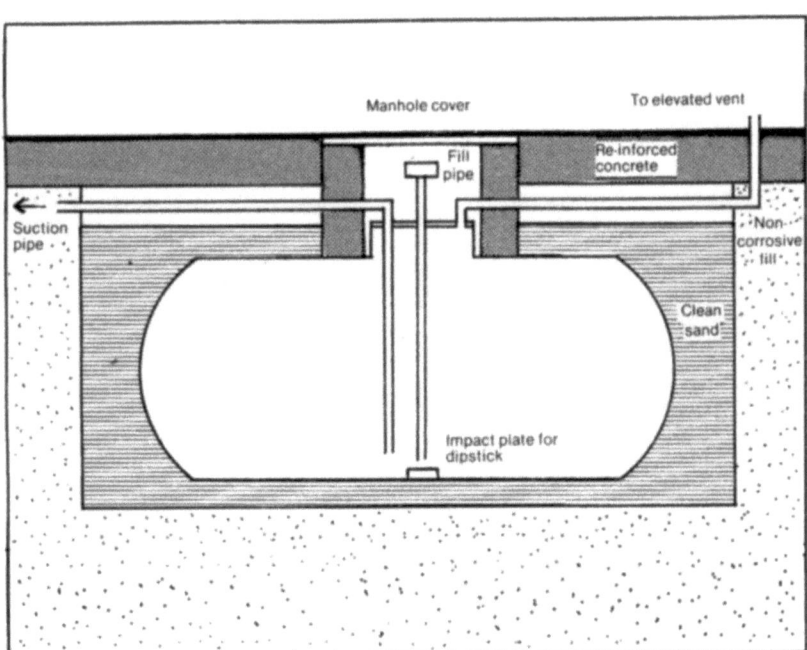

and associated pipework and carefully documented stock checks taken to detect any product loss. For larger above ground petrol tanks the licensing authority will require that tanks are located at a safe distance from other tanks, from buildings, property boundaries and from filling points and other sources of ignition. A minimum distance of 6 metres is desirable for small tanks, although this will increase for larger tanks, while very small tanks can be grouped and treated as one larger tank. Detailed recommendations for the siting of petrol and other product tanks at oil refineries and storage installations are contained in national codes and in oil company design manuals.[2,9]

BUNDING

A normal requirement for licensed petrol storage tanks is that individual tanks or groups of small tanks are enclosed by firewalls or bunds providing sufficient capacity to contain the contents of the tank or of the largest tank when a group of tanks is enclosed. *(See figure 24)* These bunds were intended initially to contain any fire and to provide protection for fire fighting personnel and contain fire fighting water as well as spilled fuel. They are increasingly regarded as desirable means of avoiding pollution and even as essential in sensitive water supply areas, since provision must be made to remove water from the bunded areas either by

Figure 24: *Horizontal cylindrical storage tank*

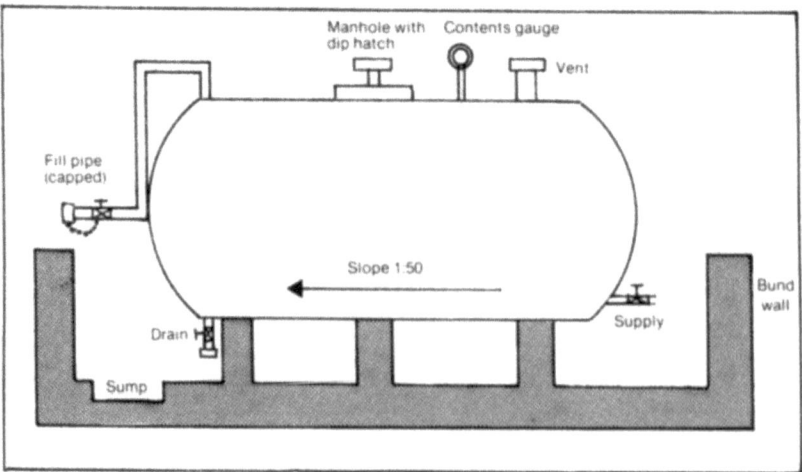

pumping or through valves which are normally closed so that the bunding provides a ready means of controlling the flow of storm water to oil interceptors. Despite the absence of legislative control for products other than Class I products, bunds are generally constructed around all large storage tanks, and planning authorities, in consultation with river and fire authorities, may request them. In areas where ground water is used for domestic supplies or where porous soil predominates, it may be necessary to provide clay, bituminous sand or other non-porous linings to bunded areas to safeguard against water contamination.

For small domestic tanks, the area under the tank should preferably be concreted and surrounded by a low curb or retaining wall and a drainage sump provided from which rain water and any spilled oil can be removed by hand pump or bailing. (*See figure 25*) At small industrial and commercial sites, good housekeeping is also facilitated by concreting around the storage area and, areas where oil handling may result in any oil spillage. This area should be provided with a drainage sump from which water can be removed under control and any oil retained for collection and safe disposal. Where larger quantities of oil are stored, the water from tank areas and other locations where oil spills may occur should be routed to oil interceptors before being discharged to drains or watercourses. In the case of petrol service stations this is achieved by having a low concrete curb round the filling area and directing rainwater to an oil interceptor. At oil company storage installa-

Figure 25: *Rectangular domestic storage tank*

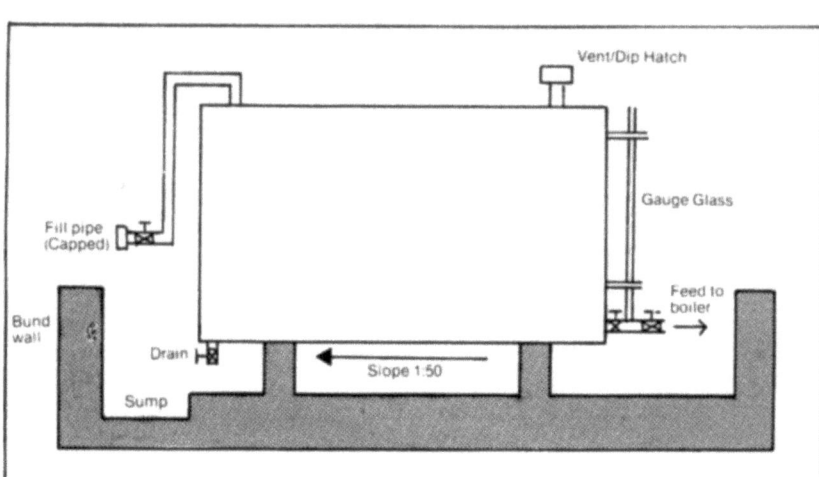

tions all storm water which may be contaminated by oil is directed to oil/water separators from which oil is recovered before the water is released.

OIL SPILL PREVENTION

Most oil spill incidents result from human errors and carelessness; very few are caused by the failure of properly designed facilities and equipment. One survey shows that all but 50 of the 461 incidents considered could have been prevented by reasonable precautions and 35 of the 50 were the result of accidents involving vehicles.[11] Clearly the definition of individual responsibility and the provision of adequate training are essential points to consider. The same survey suggests that oil spills are more likely to occur at user premises than at oil storage installations. This would suggest that the accidents result from the fact that oil stock control represents only a small part of the workload of the storekeeper or boiler operator at user premises but is the fulltime occupation of most employees at a storage installation. A number of steps need to be taken in addition to ensuring that the correct type of tank has been built to the correct specification in the best location and that suitable pipework and fittings have been installed. These should include the nomination of a person or persons to be responsible for ordering and receiving oil and for periodically checking the oil stocks and the satisfactory condition of all oil handling equipment.

For most domestic installations the continuity of supply is provided by the supplying oil company or their local agents on a routine basis. Deliveries are often made in the absence of the occupier and the driver is solely responsible for ensuring that no spillage occurs. The driver should be assisted by the use of an automatic cut-off inserted into the receiving tank and operating on the vehicle delivery pump, or by the delivery being through accurate meters on the vehicle so that only the quantity of fuel is delivered for which ullage exists prior to the commencement of delivery. Vehicles used for such driver attended deliveries should carry sawdust or other oil absorbent materials to contain any minor spills that may result from the disconnection of hoses or other incidents.

At commercial and industrial premises there is likely to be more than one grade of oil involved, larger quantities of oil will be

Plate 23: *Crude oil storage tanks on Widdie Island, Bantry Bay, Eire; note the high bund. The water in the foreground is a catchment pond for segregated ballast water.*

stored and several people may be concerned. The importance of clearly marking each tank and delivery line connection point with the grade of oil and tank to which the line is connected is obvious as also is the need to ensure that everyone concerned with the acceptance of oil deliveries is aware of the stock position at all times. Serious oil spills have resulted from deliveries being made into an already full tank instead into an adjacent empty tank or because previously available ullage had been filled by an earlier delivery without the knowledge of the person accepting the second delivery. If such incidents are to be avoided — as indeed they should — tank dips should be checked immediately prior to a delivery being accepted and all hose connections checked to ensure deliveries are correctly routed.

Stock security is another important point if spills are to be avoided. Cases have been reported of animals gaining access to tanks and turning valves or breaking pipework thus causing spills. Cases of temporary feed lines to equipment being made with unsuitable hose which parted whilst unattended are also known. More common, but still easily avoided, are the pollution incidents which result from the storage of waste. One serious incident resulted from the earth bund retaining a large lagoon of waste oil collapsing after heavy rain and releasing the oil to pollute a lake and river system causing considerable damage to fish stocks. Waste oil stored in light metal containers which are subsequently left to rust releasing their contents into an adjacent ditch or open drain are another common cause of pollution. Such incidents can be avoided by storing waste oil in sound steel drums or preferably in a purpose built tank, and arranging for the oil to be collected regularly by an authorised waste oil disposal contractor complying with waste disposal regulations which in the United Kingdom are contained in the Deposit of Poisonous Wastes Act and in an EEC Directive.[12,13] Where the quantity of waste oil does not justify the installation of a special storage tank, arrangements should be made to store filled containers on a concrete slab with a retaining curb.

Unauthorised interference with tank valves and other forms of vandalism have also been reported as causes of oil spills. Where such risks exist consideration should be given to more effective fencing and to the use of locks to secure valves and screw caps on lines adjacent to property boundaries.

After every care has been taken with the design, construction and operation of a storage facility, some incident may occur which will result in an oil spill. For this reason provision should be made for the containment of such spills as may occur and to minimise the quantity of oil involved. The provision of bunds for larger tanks and of curbs around concreted areas under smaller tanks has already been discussed, as has the question of controlled drainage from these enclosures. Other areas where oil leaks or spills may occur include oil loading and handling bays, parking and servicing areas for fleets of heavy vehicles. Such areas should be paved and drained to oil interceptors in the same way that the forecourts of petrol service stations are protected. The design and operation of interceptors is dealt with elsewhere in this volume. They are only effective if they are correctly maintained and oil is not allowed to accumulate until the retention capacity is exceeded and then washed into the watercourse they are designed to protect.

Reference has also been made to the installation of valves as close as possible to tanks. This minimises the risk of spillage due to a line being damaged, but is only effective if the valve is closed. The provision of a non-return valve on filling lines which drain away from the tank ensures that if a line is damaged during filling operations, and therefore when the tank valve is open, oil will not flow back from the tank.

Containment considerations should also include the provision of some capability to prevent oil spills entering sewers or surface drains. Sacks of sawdust or other oil absorbent material, sheets of plastic suitable for covering drains and held in place with sand or earth and a supply of open topped drums, to contain any recovered oil and the oily absorbent, are items which should be available as first aid pollution prevention equipment at all but the smallest oil storage facilities.

Finally, and in case all the foregoing measures fail, all owners, occupiers and operators of storage facilities should know what action to take in an oil spill emergency. A document on this subject has been prepared jointly by the oil and water industries and the UK government departments concerned.[14] It suggests that the police, fire and local river authorities be contacted immediately and includes a chart showing how reports of any incident may reach these emergency services.

REFERENCES
1   The Petroleum (Consolidation) Act 1928 is the authority under which storage regulations are issued.
2   Home Office Model Code of Principles of Construction and Licensing

Conditions. Part I The Storage of Petrol in Cans, Drums and other containers; Petroleum-Spirit Filling Stations, and Part II Distributing Depots and Major Installations. H.M.S.O.

3 British Standard 1515: Fusion welded pressure vessels for use in the chemical, petroleum and allied industries. British Standards Institution.

4 Institute of Petroleum Model Code of Safe Practice, Part 9 Liquefied Petroleum Gases 1967. Institute of Petroleum, London.

5 British Standard 799; Oil Burning Equipment, Part 2 1964, Oil storage tanks up to 750 gallons; Part 5 1975, Oil Storage Tanks. British Standards Institution.

6 British Standard 2594: 1975 Carbon steel welded horizontal cylindrical storage tanks. British Standards Institution.

7 British Standard 2654: 1973 Vertical steel welded storage tanks with butt-welded shells for the petroleum industry. British Standards Institution.

8 British Standard 3792: 1964 Recommendations for the installation of automatic liquid level and temperature measuring instruments on storage tanks. British Standards Institution.

9 Institute of Petroleum Model Code of Safe Practice, Part 2 Marketing. 1978.

10 Institute of Petroleum Model Code of Safe Practice, Part 10 Storage and Piped Distribution of Heating Oil, 1967.

11 R.G. Toms, *Water Pollution by Oil*, (Institute of Petroleum, London, 1971).

12 Deposit of Poisonous Wastes Act 1972.

13 EEC Directive 75/439, 16 June 1975.

14 *Inland Oil Spills, Emergency Procedures and Action* (Institute of Petroleum, London, 1978).

11

Discharges from industrial plants and the like into sewers, rivers and the sea

G.F. Oldham
*Effluent Adviser,
The British Petroleum Company Limited*

So far, this book has discussed the various ways in which oil can be spilled at all stages between the oil well and the final consumption of the product. It will have been seen that the majority of spills are the result of the failure of equipment or of human error. However, as pointed out in the first chapter, neglecting oil from various sources carried by the atmosphere, about one third of the oil entering the sea originates from urban run off, factories, refineries, general industrial run off and untreated domestic sewage, either discharged directly into the sea or from rivers. Most of this can be prevented by installing proper separator equipment and by the exercise of due care at all times. This chapter discusses the types of equipment and treatment systems needed to deal with oily effluents of all kinds including those from complex oil refineries. For many less sophisticated sites, of course, the simpler types of treatment will be adequate.

Oily discharges arise from a number of sources. The principal are:
    Oil refineries;
    Petrochemical works;
    Factories;
    Oil Storage sites, depots and stores;
    Petrol filling stations;
    Engineering workshops;
    Steel works.

In addition there are discharges of oil into drains and on to the land by people servicing their own motor cars and having to dispose of waste oil. This latter problem can be dealt with by the provision of collecting points for the oil and the application of local bye-laws to ban discharges of this kind.

Sources and types of oily effluents

Oily effluents which may or will need treatment include:
    Process effluents, wash waters, oil condensates, etc.
    Drainage water from crude oil and product tanks;
    Production water arriving at an oil storage site or refinery terminal along with crude oil from a production site;
    Ballast water from crude oil or product shipment;
    Water from washing vehicles and hard standing areas;
    Water pumped from fixed and variable water beds in oil and oil/product storage caverns;
    Oily or potentially oily rainwater.
    Oily cooling water and blow-down from circulatory systems.

'Housekeeping' and re-use of effluents

The more one can reduce the flowrate of oily effluent to be treated and the more one can eliminate difficult or emulsified effluents, the small, simpler and cheaper will be the treatment plant. These goals can often be achieved by good 'housekeeping' and re-use of effluents.

Once-through cooling water systems which are susceptible to oil leaks can be replaced by closed circuit recirculation systems with cooling towers where only the blow-down from the system needs treatment. (The blow-down is the water which must be bled out of the system to prevent the mineral salts in the make-up water being concentrated to a level where they form scales and deposits).

Oily rainwater after storage in a tank or pond is often acceptable as make up to a closed circuit cooling system if the oil has been effectively skimmed. Sometimes steam condensate which is, or is likely to be, oily or contaminated is dumped into a closed circuit cooling system. The addition of rain water or condensate, being free from calcium salts, also reduces the flowrate of the blow-down. Re-use in firewater mains may also be considered.

Oily rainwater can be reduced or eliminated by erecting roofs over those areas which are or could be oily such as pumps, machinery areas, drain cocks, garage areas, parking or standing areas, maintenance bays and places where oil pipe or hose connections are made and remade.

Segregation

Apart from oily effluents, most, if not all, sites have effluents which are not oily. These include:
  Rainwater from undeveloped areas, clean areas and roofs;
  Regeneration effluents from ion exchange vessels for boiler water treatment;
  Blow-down from boilers;
  Cooling water which has not been or cannot be contaminated by oil.

These should be collected in a separate drainage system which by-passes the oily effluent treatment system.

Effluent streams containing significant quantities of detergents or emulsified oil should be kept out of the main oily effluent as they will probably need special treatment and could otherwise emulsify the oil in the main effluent.

Any oily spent soda which cannot be re-used should be neutralised before mixing with the main oily effluent.

Some inland oil refineries in Germany and France have tight limitations on the flowrate of treated effluent which may be discharged into a river and/or the treated effluent has to be piped a long way to the nearest receiving watercourse. In these cases, the 'salty' effluents (desalter water, domestic sewage, boiler blow-down, cooling tower blow-down) are segregated and treated separately for discharge, while the non-salty effluents (oily rainwater, oily condensates, pump cooling water, wash waters and product tank drainage) are treated to a quality suitable for make-up to the cooling system.

Where a local authority foul sewer or 'main drain' is available, domestic sewage, bath-house water, laundry effluent and laboratory effluent (after passage through a simple oil trap), should be segregated and discharged to the foul sewer, as they would otherwise seriously interfere with treatment of the main oily effluent stream. Alternatively, they may be treated separately by septic tank or similar system.

Where such a main drainage system is available, the local authority will normally accept trade effluent, at a calculated charge, provided that their own treatment plant has the capacity to accept it and provided that it meets certain quality standards designed to safeguard the fabric and good operation of the sewers, any maintenance personnel working in them, their own treatment plant and the quality of the final receiving water. For an oily effluent, this will almost certainly require some pretreatment.

The rest of this chapter lists treatments which may be selected for either pretreatment for discharge to a local authority main sewer or for complete treatment for discharge direct to a receiving water such as a river, canal, estuary or the sea.

TREATMENT OF OILY EFFLUENT

General

Oil is almost insoluble in water and usually has a lower specific gravity. Most of the oil present in oily effluents can therefore be separated by gravity and skimmed off provided it is not emulsified or too finely divided. To avoid the formation of such finely divided droplets, it is most important to avoid pumping oily streams before they have passed through the main gravity separation, especially if there are traces of detergent present.

Oily effluents should therefore be allowed to drain by gravity to the gravity separator, which is therefore normally at the lowest part of the site or recessed in the ground, or both.

Depending on whether it is possible to discharge to a local authority sewer or to a watercourse (river, estuary or the sea), the effluent may need treating by either gravity alone, or gravity plus enhanced gravity or other physical methods, or may need treating by biological means as well, or may even in extreme cases need some final polishing treatment.

Following the standard terminology of classical domestic sewage treatment, gravity and similar treatment is known as 'primary' treatment, biological treatment is known as 'secondary' treatment and any further polishing treatments are known as 'tertiary' treatment. With oily effluents, it is common to have a polishing stage to the primary treatment and this is sometimes called an 'intermediate' treatment.

Primary treatments

Oily effluents are almost invariably passed first through a gravity separator such as a simple rectangular basin, preferably of American Petroleum Institute (API) design, or a circular clarifier or a plate separator.

API separator

This is described in the API Manual.[1] The principle is to slow down the oily water entering the inlet bay and to allow it to pass slowly as a smooth undisturbed flow along the length of the main bays (at least two similar parallel bays). Droplets of oil rise towards the surface at a rate depending on their density difference with the water, their size and the viscosity of the water (which depends on the temperature). For a droplet to reach the surface, its rise rate must be equal to, or greater than, the 'overflow rate', i.e. the flow rate divided by the plan-view surface area of the main bays and it must not be caught by downward movements of water due to turbulence caused by, for example, residual kinetic energy from the inflowing oily water or obstructions in the main bays.

The efficiency of separation therefore depends mainly on the area of the main separating bays (the depth and 'retention time' therefore have relatively little effect so long as they are within reasonable limits) and the efficiency of destroying the kinetic energy of the incoming water before it passes into the main bays. This is best done by separating the inlet bay from the main bays by a number of vertical concrete pillars, spaced fairly closely. The kinetic energy is then dissipated mainly in the inlet bay. As much of the oil in the incoming oily water will be in the form of very large globules that separate very easily, this will separate in the inlet bay and can be skimmed from there. The medium sized globules rise to the surface in the main bays and are also skimmed off.

The water, still containing the smallest globules, emulsified oil, soluble materials and suspended matter with an effective density close to that of water, passes under an inverted weir (which holds back the oil which has risen to the surface) and then over a weir (which maintains the level in the separator) into the outfall bay (from which it can pass on by gravity or be pumped to the next treatment stage).

Similarly, solids with a density greater than the water will settle out at the bottom or be carried through in the water depending on their settling rate. Oil-coated solids and oil droplets with solid matter attached can have an effective density close to that of the water and can be carried through, even though their size is appreciable.

Assuming ideal conditions, it is possible to calculate the size of API separator required to separate down to any particular globule size. The usual size is 150 micron (0.15 mm).[1]

As mentioned above, it is important to avoid downward movements or turbulence in the main separating bays and it is therefore important to avoid mechanical apparatus in the water which could cause eddies. This means that the design of equipment to skim surface oil and to remove from time to time the solids which settle at the bottom needs careful thought. The type of flight scraper which passes continuously one way through the surface oil then dips down through the water and scrapes the bottom solids in the opposite direction, not only causes much turbulence but sheds much oil or oily solids into the cleaned water just passing out of the separator. This type is therefore best avoided.

Flight skimmers and scrapers, if used, should be separate, not combined, and of a design which projects into the water as little as possible. They should only be operated when necessary.

The actual removal of oil from the surface is either by skimming into a rotatable pipe partially immersed in the surface with a longitudinal slot, either at the inlet or outlet ends of each main bay, or both, or by rotating disc skimmers. The slotted pipe method skims a large quantity of water with the oil which is then allowed to separate in a skimmed oil sump. The water layer is then pumped back to the inlet to the API separator and the oil layer is pumped to the recovered oil tank.

The rotating disc type has the advantage that it recovers dry oil and can be left running whereas the slotted pipe requires intermittent manual operation. A number of discs on an axle rotate

Figure 26: *Longitudinal section of an API type separator.*

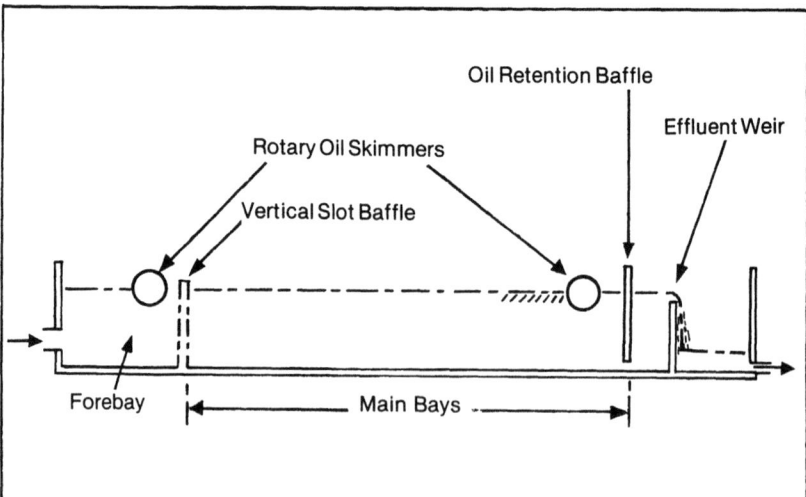

and dip through the oil, picking up the oil. After rising out of the water and before dipping down again, the oil is scraped off into pouches which drain the dry oil away to direct recovery. Alternative disc drive systems are available to suit safety and other requirements.

Other than effluent quality improvements possible with disc skimmers, the ability to operate continuously may obviate the need for covered separators (see later). As opposed to the intermittant operation of slotted pipe weirs, continuously operating disc skimmers do not permit build-up of oil layers with the consequential release of hydrocarbon vapours.

The solids which settle to the bottom of an API separator vary from fairly coarse heavy material such as sand and rust to very fine material derived from fine mud solids or precipitated chalk which can have a consistency more like potato soup (which may not be detected by a dip rod). If the quantity of coarse bottom solids is expected to be large, it might be preferable to have a bottom flight scraper with a recess in the bottom at one end from which the sludge can be pumped. For normal quantities such as in a typical oil refinery or.petrochemical factory it is probably best to remove the fine sludge from time to time using a vacuum lorry and to remove the heavy solids every few years, normally after draining each bay.

API separators are necessarily large but they have a number of important advantages. They can accept widely varying proportions and varieties of oil and solids including viscous, sticky or waxy oil and can retain very large quantities of oil after major accidents. Many sites treat the large volume of water always present as a useful emergency firefighting supply.

The question is sometimes raised of hydrocarbon vapour loss from the oil on the surface and to what extent it is practicable to cover the surface. In the main bays, anything floating on the surface will prevent skimming of the oil, making matters worse and a roof, besides making skimming difficult, is a serious explosion and fire risk. With adequate size of inlet bay, most of the oil will separate there and it is usually practicable to have a floating raft which will reduce the hydrocarbon loss to less than one tenth.

Circular separators or clarifiers
These have central inflow of oily water.[1] Clarified water passes out under and over peripheral weirs. The linear velocity of the water thus decreases radially. The even distribution of the water flow is not easy. In general, this type has proved more useful and

much more used for separating solids by sedimentation, e.g. after flocculation or in the activated sludge process (see below). The sludge is removed with a rotating scraper on the conical bottom and pumped away.

Plate separators
These are in effect a large number of shallow separators stacked above each other and operating in parallel. They consist of plate packs with the plates spaced 4" (100 mm), or usually nowadays less, apart. The oil droplets therefore only have to rise this short distance before reaching the next surface above. The solids settle downwards similarly. The plates are sloped so that the oil can rise to the top of the pack and sludges may slide downwards. Because of the compact design it is practicable to design for separation of oil droplets down to 60 micron diameter (0.06 mm). To avoid corrosion problems the plates are normally made of plastic.

The advantages, besides compactness, are that the whole pack can be enclosed and operated full of liquid, preventing any vapour loss, or they can be enclosed in a pressure vessel. Oil can be removed on level control.

The disadvantages are that they can easily be blocked by viscous or waxy oil or thick sludge, they have very little capacity for retaining large 'slugs' of oil following accidents and the large plate surface area has a strong tendency to grow bacterial films causing slightly bacterial oily sludges which need steam treatment to break them down to handleable recovered oil. For these reasons they usually need cleaning at fairly frequent intervals and provision for this should be included when they are installed.

They are particularly appropriate for oil production site formation water, which is usually always warm and the oil fluid, and where space and weight are severely limited.

There are a number of physical modifications to plate shape, direction of slope and attachments to facilitate removal of oil from the plates as in the parallel plate interceptor (PPI), tilted plate separator (TPS), corrugated plate separator (CPS), etc.

Three-chamber petrol interceptor
This is a simple structure of brick or concrete, consisting of three small chambers in series. The water enters the first through a pipe and is led successively through the others via pipes coming up from the lower region of each previous chamber. It is designed for trapping spillages of petrol or similar very easily separated light liquids from small water flows. The first chamber fills with petrol

first, then any further petrol gets carried through to the second chamber, etc, so a simple check on how far the petrol has progressed through tells one how soon it needs emptying.

As the design of inlet and outlet of each chamber is usually very simple and against the principles of good API separator design, the efficiency is extremely poor and they cannot be considered as oil/water separators.

'Intermediate' treatments

These are designed to remove some or all of the remaining suspended oil in the water leaving the primary gravity separator. This suspended oil consists of small globules of about 150 micron diameter or less, fine oily solids, larger oil globules with attached solids and any emulsified oil present.

Coalescers

Coalescers are normally coarse filters (often granular) or fibre-packed cartridge which are designed to facilitate coalescence of small oil globules so that when the oil finally passes out of the packing it is in the form of large globules which separate easily. For continuous operation it is usually most convenient for the fine solids present to pass through and out with the coalesced oil. However, coalescers can act as filters for the solids present and tend to become blocked unless the solids can be backwashed. Emulsions normally pass through most designs of coalescer.

Flocculation

Finely divided oil and solids and many emulsions can be effectively treated by flocculation with precipitates of the hydroxides of aluminium or iron. The incoming liquid is mixed either with a solution of aluminium sulphate or chloride and the pH adjusted to about 6 – 6.5, or with a solution of ferric sulphate or chloride and the pH adjusted to about 8. (The correct pH is very important and the optimum value is found by jar tests in the laboratory on the actual effluent before commissioning.) The hydroxide floc is then allowed to form, typically by stirring for a few minutes followed by slower stirring or agitation for another, say, 15 or 20 minutes. Fine suspended matter and much of any emulsion present is absorbed and adsorbed onto the flocs, which should be relatively large and easily settleable. Flocculation is normally followed by sedimentation of the floc and removal of the sedimented sludge in a clarifier or may be followed by dissolved air flotation (see below).

In some cases flocculation may be more effective with the further addition of a suitable polyelectrolyte.

Flocculation followed by sedimentation is a very effective means of removing practically all the suspended matter and much of many emulsions. It does, however, require the constant addition of chemicals, preferably with proportional dosing, pH control and disposal of the oily sludge formed. The latter is often by thickening in a 'thickener' (a circular tank with a slowly rotating paddle) followed by dewatering (by vacuum filter or centrifuge) and final disposal by incineration or removal by contractor for incineration or other approved disposal method.

Flotation

The principle of flotation is that any suspended particle or globule to which an air bubble can be attached and remain attached will rise relatively rapidly to the surface from where it can be skimmed and removed for further treatment (such as thickening, coalescence, disposal or recirculation). The effectiveness of this type of treatment depends on the proportion of suspended particles and globules to which an air bubble can be successfully attached and remain attached during their rise and skimming. In general, bubbles attach themselves more easily to solid particles than oil globules. Coalescence within a flotation unit is highly desirable.

There are two basic forms of flotation — dissolved air flotation (DAF) and induced air flotation. Either form may use inert gas (e.g. nitrogen) or hydrocarbon gas instead of air where circumstances dictate, e.g. where there is an explosion risk with air as on oil production platforms. Flotation has often been used in conjunction with activated sludge (see below).

Dissolved air flotation. In this form, a proportion, up to about half, of the total water leaving the unit is passed through a recirculation pump. Air is also fed into the suction of this pump. The mixture of water and air under pressure is passed through a small vessel or packed bed where the water becomes saturated with air at the particular pressure. This saturated water is then recirculated via a pressure reduction device to the inlet of the unit where it mixes with the incoming water. The reduction in pressure causes the air to come out of solution. This air tends to form small bubbles on any nuclei present.

The effectiveness of this treatment is increased by adding organic polyelectrolytes and these when used alone allow recirculation of the skimmed foam to the inlet of the primary gravity separator.

The effectiveness can be markedly improved by previously flocculating the suspended matter, e.g. with aluminium or iron salts (either alone or in conjunction with polyelectrolyte) but then the skimmed foam contains a metal hydroxide sludge which must not be returned to the primary gravity separator (as this would tend to flocculate the larger oil globules and prevent their coalescence and recovery) but must be disposed of as above (see flocculation). In practice, some, occasionally most, of the floc particles are not floated but pass through and may settle out in the next stage.

Induced air flotation. In this form, one or several rotors under the water surface with hollow vertical shafts, suck or induce air into the water and rapidly disperse it generating much foaming. Normally only polyelectrolytes are dosed (aluminium or iron hydroxide flocs would be broken down with the violent agitation) and the suspended matter tends to become coalesced and floated to the surface from where it is skimmed with slowly rotating paddles and recirculated to the inlet of the primary gravity stage.

This type of flotation is much used in conjunction with plate separators in pressure vessels as a reltively compact, lightweight and continuously operating treatment for production water on oil production platforms.

Filters

Filters can be very effective for removing virtually all suspended matter, both oil and solids, but not true emulsions. For continuous operation they must be backwashable and, for this reason, granular filters are normally employed. Engineers commonly refer to these as 'sand' filters but the filtering medium may be silica sand, garnet sand, crushed anthracite or various combinations of these in layered beds. They are backwashed with various combinations of water and air. To allow continuous treatment, a bank of filters in parallel is used so that one can be backwashed while the rest are filtering the full flow of oily water. Backwashed oil and solids may be recirculated to the inlet of the primary gravity separation stage.

For particular cases, precoat or cartridge filters may be used but these need a regular supply and disposal of used precoat material or cartridges.

Because of their ability to remove virtually all the suspended oil present, there has been a steadily increasing use of filters combined with biological percolating filters for treating refinery and similar oily effluents.[3]

Secondary (biological) treatment

Where a treated effluent is discharged to a river and in a growing proportion of instances where it is discharged to an estuary or the sea, the authority which permits the discharge will usually limit the biochemical oxygen demand (BOD). This is the capacity of the effluent for leading to a removal of dissolved oxygen in the receiving water over a period of several days. This is caused by the natural bacteria present in receiving water utilising organic and inorganic matter present in the effluent as a source of food and energy and allowing them to use up some or all of the dissolved oxygen in the water. This could lead to the death of fish and other animals and, in severe cases, to anaerobic conditions giving rise to hydrogen sulphide emission. The BOD is measured in a simulated natural environment in a bottle over a period of five days.

The logical, easiest and cheapest way of reducing the BOD of the effluent is to design for this natural bacterial process to be carried out on the effluent before discharge and in ideal conditions with adequate supply of oxygen. In reducing the BOD of the effluent, the bacteria also consume various unwanted chemicals present such as phenols, sulphide etc., and convert them partially to increased body mass and partially to carbon dioxide, water and sulphate.

Suspended oil is only partially destroyed by the bacteria in the time scale of practicable treatment and the residual oil, by its smothering effect on the bacteria, can cause substantial reduction in efficiency and a final disposal problem unless suitable precautions are taken.

Most analytical methods for oil also extract and measure some of the dissolved organic matter. This is part of the material giving rise to the BOD and is effectively reduced by biological treatment. There are two main types of biological treatment — one in which the bacteria grow in a film on surfaces which are part of the equipment and the effluent to be treated percolates over or between these surface films. These are known as percolating filters. In the other the bacteria are in suspension in the effluent to be treated in an aeration zone and are later settled out and returned to the aeration zone; this the activated sludge process.

Percolating Filtration

Percolating filters can be in the following forms:
> Trickling filters with random fill stone filling or ordered plastic sheet packing or random fill plastic packing or layered combinations of packings;
> Rotating discs or cages of random fill.

In all of these, a slimy film of bacteria, attached to each other by a gelatinous secretion, grows on the surfaces exposed to the effluent, feeding on and removing the substances which give rise to the BOD. As the film grows thicker, fragments become detached, slough off and come out of the percolating filter in suspension in the treated water and are known as humus. The concentration varies enormously (depending on changes in flowrate, temperature, pH, etc.) but the average concentration is approximately proportional to the BOD of the water at the inlet. If the average concentration of humus is more than about 30 mg/l (parts per million, weight/volume) it is normally settled out in a clarifier and removed.

Percolating filters have the important advantage of being completely automatic in operation and require no operator attention and practically no maintenance. They are also very resistant to 'shock loads', i.e. high flowrates, moderately high temperatures and pH values and recover by themselves, and very quickly, from severe effects of temperature, acid, alkali and toxic substances.

They have proved very practicable and suitable for the dissolved impurities in most oily effluents such as from oil refineries and petrochemical factories. However, any suspended (i.e. insoluble) oil present has two very important effects. Initially, the biological efficiency is severely reduced by the smothering effect of the oil. Furthermore, suspended oil readily adsorbs onto the biological film and gives it a more viscous and adhesive quality. For a short term dose of suspended oil, this will slowly slough off bringing most of the oil out still adsorbed on the humus solids. However, for a regular application of quite small concentrations of suspended oil, the oily film builds up and can cause eventual blockage of the filter. It is therefore essential to remove virtually all the suspended oil before the percolating filter. Most intermediate treatments are not sufficiently effective or sufficiently reliably effective and some attempted combinations of, for example, flotation and percolating filtration have led to blockage of the percolating filter. However, granular filters followed by percolating filters have been found a satisfactory, powerful and economic

Mass Transfer Ltd
Filterpak

combination.[3] Furthermore, the BOD_5 of filtered refinery effluent, for example, is normally sufficiently low for the humus solids in suspension to be comfortably less than 30 mg/1 after averaging out, and the usual humus removal stage can be omitted, saving cost and operational requirements, and replaced by a small lagoon operating under equilibrium conditions (see below, section on lagoons).

Waste waters are applied to the trickling forms of percolating filter via a reaction-driven rotary distributor over the medium contained in a circular tank through which air rises by natural thermal convection. Other forms of distributor are sometimes used.

The stone-filled trickling form has been used for the treatment of domestic sewage and other effluents since the 1890s. Since the 1950s, ordered-sheet and later random-fill plastic packings have been much used. Although the packings are more expensive per cubic metre than crushed stone, they are lighter in weight and due to their purpose design they are more efficient thus allowing smaller size, lighter foundations and greater height to diameter ratio. The plastic used is normally polypropylene but PVC and other plastics have also been used.

Sometimes percolating filters are made in the form of rotating discs, or cages of random fill media, which rotate on axles and dip in and out of the effluent, lift some of the liquid up on the surface and allow it to percolate down again. Their mode of operation is very much the same as the more usual trickling form. They are particularly suitable for effluents whose flowrate ceases for considerable periods, e.g. overnight. When this occurs with the trickling form, recirculation is used.

Since about 1974, the use of percolating filters for oil refinery and petrochemical factory effluents has been rapidly superceding the use of activated sludge mainly on the grounds of low installed capital cost, much lower operating costs, simplicity and reliability, and absence of operators.

Activated sludge
This system, which was developed in Manchester, England, about 1913, dispenses with the surface on which the bacterial film grows, and the bacteria are in suspension in the effluent in the form of an active suspended biological sludge. This is aerated, using bottom diffusers or surface aerators, in one or several aeration basins in series. After sufficient retention time, the effluent, still containing the suspended activated sludge, passes to a clarifier from which nearly all the sludge must be pumped back

Plate 24: *A biological percolating filter showing rotary distributor.*

to the inlet to the aeration basin. The growth of the active sludge is balanced by that lost in the outflow from the clarifier and a controlled bleed-off of the excess sludge from the system.

The efficiency of the process is closely related to the quantity, i.e. concentration of active sludge in the aeration basin. As all this is in suspension, it is vital not to lose more than traces from the final clarifier. This can occur if the flowrate exceeds the design flowrate of the clarifier or if, for example, dense cold salt ballast water flows into a clarifier containing sludge suspended in warm fresh water, or if 'bulking' occurs preventing adequate settling in the clarifier, or if the inflowing BOD is low and insufficient to maintain at least as much growth of sludge as that lost in the outflow from the clarifier. Also, the sludge is subjected to any shock load such as temperature, pH, toxic substances, etc. As all the bacteria are immersed in the liquid, they may thus be rendered inactive for sufficient time for them to be lost from the system, or the whole sludge may be killed. Regrowth of sufficient activated sludge may take several weeks.

When used for oil refinery or petrochemical effluents, it is common to add iron salts with the inflowing water together with sufficient alkali to precipitate a floc of ferric hydroxide. This absorbs the residual suspended oil and provides a convenient medium for the bacteria to grow on.

Where any suspended oil enters the activated sludge stage, the excess sludge is oily. It is then normally thickened, dewatered and incinerated as described above in the section on flocculation.

Activated sludge plants usually have a responsible person in charge and operators on shift.

The combination of flotation followed by activated sludge has often been used and one plant treating a combined oil refinery and petrochemical waste water with a high BOD has used a combination of sand filtration followed by activated sludge.

Where the inflowing BOD is below about 100 mg/1 or the temperature of the effluent to be treated could rise above about $40°C$, the activated sludge process is not suitable. In cases such as these and for small and medium sized plants, the activated sludge process has tended to be superceded by percolating filtration.

Lagoons
These are shallow ponds or lakes through which partially treated effluent is passed and the word has been used in two main senses.

If, say, a weak refinery effluent of BOD say, between 50 and 100 mg/1 is passed through a large shallow lake with a retention time of the order of a month, a slow biological action will develop and may improve the quality of the effluent to within the requirements of the authorities. Aeration of the water can be by floating aerators or in some cases diffusion from the atmosphere may be adequate. These lagoons are commoner in warm, sunny climates such as California. Where possible, they are, of course, very cheap to operate. However, the value of the land is usually prohibitive and the degree of treatment is not always adequate. There is also a slow fouling up with oily sludge which may eventually need removal and disposal.

The other use of the word is for a much smaller shallow pond placed after a percolating filter. This has two main functions. It allows the concentration of humus solids to average out in the treated effluent (and also the BOD etc.). Furthermore, although the humus solids pass out in the final effluent, their relatively long residence time under starvation conditions allows the bacteria to lose much of their body weight by respiration thus giving a lower final suspended solids concentration.

Tertiary treatment

Where a particularly high quality of treated effluent is required, e.g. for discharge to a small river or exceptionally clean or important receiving water, a tertiary treatment is sometimes required. The most usual form is to remove as much as possible of the residual suspended solids after the secondary (biological) stage. These account for much of the residual BOD (and also measured oil content).

They can be removed by filtration or flocculation and sedimentation, or by running the effluent over and through a sufficiently large grass plot. The grass acts as a combined physical filter and biological filter. The lush grass might be suitable for grazing if two alternating fields are used.

Special treatments include ozonisation (which is useful for removing final traces of biologically resistant 'phenols' but has been known to liberate cyanide from ferro- or ferri- cyanide complexes) and activated carbon treatment (normally after filtration to remove suspended matter first). The latter is not always suitable and the cost is high. Regeneration of activated carbon is a complex process and itself can lead to environmental problems. Chlorination has been used for final sterilisation but may leave behind chlorinated residues which may be toxic. Reverse osmosis is rarely applicable because of fouling of the membrances by solids and oil.

Selecting the right treatment for an oily effluent

The main basic types of treatment stages have been surveyed above. There are also many modifications or special versions of these basic stages which have not been mentioned. They may be used separately or in various combinations depending on the range of qualities of the oily waste water to be treated and the quality required by the authority which will receive the treated effluent.

If the final effluent from the factory or installation is to go direct to the sea or to a river or smaller watercourse, it is vital to establish at a very early stage in planning, what quality will be acceptable. the present and expected uses and condition and flow-rate of the receiving water must be understood, and the permissible effluent quality assessed accordingly. Effluent specification quality will be stated by, or agreed with, the controlling authorities, e.g. Water Authority. The necessary treatment processes cannot be defined and engineered without complete specification of both the raw effluent and the final required quality.

The operator of a site producing an oily waste water which requires treatment will have to select a treatment or combination of treatments which will treat his waste water to the required quality with the optimum combination of low capital cost, low operating cost, low maintenance cost, reliability and minimum operator attention. This requires considerable experience and each particular waste water needs careful study to arrive at the optimum solution. Some general comments are given below.

Firstly, one must determine the main parameters of the waste water. These include:

Flowrate — average, minimum and especially the maximum. (This will normally be during the heaviest rainfall combined with maximum process effluent and it is most important not to underestimate it.)

Oil content — maximum concentration

Suspended solids — maximum concentration and weight of wet settleable solids in a year

BOD_5 — average, maximum and minimum and average load (kg/day)

Temperature — range

pH value — range (this is a measure of acidity or alkali)

Any other relevant substances, e.g. sulphides, phenols, ammonia.

Secondly, ascertain whether there is a local authority main sewer and under what conditions effluent may be discharged into it. This disposal route is likely to be available in towns and could apply to oil storage depots, some workshops and petrol filling stations and some smaller and medium sized factories. The oily effluent from these sites in many cases will be mainly oil in relatively clean water and might be acceptable in a local authority sewer after primary treatment only or perhaps primary plus intermediate treatments.

Emulsified oil may be a problem and this can sometimes be solved by segregation followed by disposal by contractor or perhaps treatment through disposable activated carbon cartridges or batchwise manual flocculation and sedimentation.

Steelworks have very large effluent flowrates, the impurities being mainly oil and suspended solids. A combination of API separator followed by sand filtration often gives a good quality treated effluent.

Oil refineries are usually in isolated locations and must discharge their treated effluent usually to the sea or an estuary. The flowrate usually has a maximum value in the 500-1000 m^3/h range. The usual treatment is API separator, followed by either sand filtration and percolating filtration, or flocculation and/or flotation, activated sludge and incinerator, or some similar scheme.

Petrochemical factories have effluents usually in the same broad band as oil refineries and the two are not infrequently found combined. Similar treatments are appropriate.

Plate 25: *API separator, sand filters and percolating filters in use at a refinery.*

Ballast water

Many sites receive 'dirty' (oily) ballast water from sea-going or river ships or barges. Formerly, this was treated separately from the main site effluent usually by 24 hours' static settlement in tankage and primary gravity separation, sometimes followed by a sand filtration or flotation. More and more, ballast water is being run down at as steady a rate as possible into the main site effluent treatment plant, e.g. on oil refineries.

Biological treatments have been shown to be able to accept variations in sodium chloride salinity up to full sea water strength although if the water is cold the rate of treatment will be slower.

Extreme or sudden variations should be avoided if possible, especially at sodium chloride concentrations above that of sea water.

REFERENCES

1. *Manual on Disposal of Refinery Wastes* (volume on liquid wastes), (American Petroleum Institute, Division of Refining, Washington, 1969).
2. S. L. Hobkinson, 'Effluent water treatment — a new approach', *International Environment and Safety*, June 1978, p.28.
3. G. F. Oldham, 'The BP Effluent Treatment Process' (NPRA Annual Meeting, 29 March 1977, San Francisco) G. F. Oldham, 'Oily effluent treatment by the BP Treatment Process' (Institute of Water Pollution Control Conference, 1979 (in preparation)).

12

The human factor in oil pollution

Professor W.T. Singleton, MA, DSc
Head of Applied Psychology Dept
University of Aston in Birmingham

In the glamour of the high technology systems of the oil industry it is easy to lose track of the fact that the foundation of the whole business is people. We extract, transport, refine and use oil to serve human needs. These processes are designed and conducted by skilled human operators and if something goes wrong it can only be because somewhere at some time one or more of these operators made mistakes. There is no other cause of accidents or source of pollution.

Such incidents are important because we collectively consider that neither people nor the environment should be damaged by ill-considered decisions of policy-makers or by the technical experts who are functioning on our behalf. Our attitudes and opinions are not always rational, but neither are those of the specialists. For example, specialists in the analysis of accidents develop conventions that certain incidents occurred because of human error and others did not. If an operator turns the wrong valve at the wrong time that is a human error. Yet if one looks at the former in detail it invariably turns out that it was not entirely the operator's fault, perhaps the valve was inappropriately labelled or positioned, the procedure was not clear enough, he was not experienced enough and should never have been left with that responsibility or he had been working too many hours at a stretch. All these are mistakes, but not by the person who turned the valve. In the case of the corroding valve this might have been due to an inadequate maintenance inspection, a wrong decision during the system design which incorporated a valve with an unsuitable specification, a mistake by the valve designer who used the wrong materials or going right back to the origin, a metallurgist who had insufficient understanding of metal degradation under a variety of solvents and stresses. Which of these phenomena is called a human error is very much a matter of the current fashionable convention.

Similar conventions govern the attitudes of ordinary people to untoward events in the oil industry. In western society there has been a very considerable change in attitudes over the past fifty years, largely due to the mass media. At the time of the First World War we recognised that our lives were influenced and often terminated by the decision of statesmen and arms' manufacturers but these people were remote, godlike figures different from us, not subject to our criticism so we put up with things. Now, thanks to the media we know that they are human and fallible and when they make mistakes we expect something to be done about it. Correspondingly, the senior policy makers accept the influence of public opinion and their decisions are conditioned by it.

This social and political situation may seem rather remote from incidents in the oil industry, but it is a necessary background to the understanding of what happens in practice. For example, if a diver is killed in his car travelling to Aberdeen, that is too bad but it happens to hundreds of people every day throughout the world. If the same man is killed under an oil platform in the North Sea, that is serious and is worthy of an elaborate investigation costing very large sums of money. This difference is partly a response to public opinion and partly a guilt-feeling about taking something valuable from under the sea or land and there is great sensitivity that the cost in human life or quality of life must be kept to a minimum. Governments, oil companies, shipping companies and insurance companies all seem to be subject to this rather unbalanced sensitivity. However, questioning the rationality does not reduce the effect. Accidents and pollution in the oil industry are and will be the subject of extensive study and financial investment. One approach is to postulate that we have to reduce the probability that those involved at any level will make mistakes. The background to this is theory and procedures concerned with human error.

THEORIES OF HUMAN ERROR

Each theory about behaviour makes its own contribution to the understanding of why people make mistakes. All the different approaches[1] can be summarised into three main types; human performance theory, personality and social theory and skill and systems theory.

Human performance theories

From this point of view the human operator is an information channel which can be examined to find the relationships between inputs and outputs or stimuli and responses.[2]

The first principle which emerges is stimulus-response compatibility or population stereotypes. As the name implies these are habits which certain populations have acquired and if the designer accidentally contravenes the implicit standards which the operator uses then mistakes will be that much more likely. For example, everyone expects scales to increase clock-wise, from left to right and from down to up and a scale not conforming to this will be misread as shown in figure 27. Similarly when we move a control we expect clockwise, left to right, and down to up movements to increase whatever is being controlled. Manually operated valves result in some ambiguity because a right-hand thread will usually result in a valve being closed by a clockwise movement, thus reducing some flow or pressure, which contravenes the normal stereotype. If the operator has time and attention to think about what he is doing he will get it right, but in an emergency when he tries to act quickly he may well turn it the wrong way thus accentuating the emergency. Similar principles apply to relationships between stimuli and responses, we expect a large piece of material or a pointer on a dial to move in the same way that we move the appropriate control and if it does the opposite, confusion will result. A highly ambiguous situation occurs if the

Figure 27: a. *Designs conforming to stereotypes;* b. *designs encouraging misreading.*

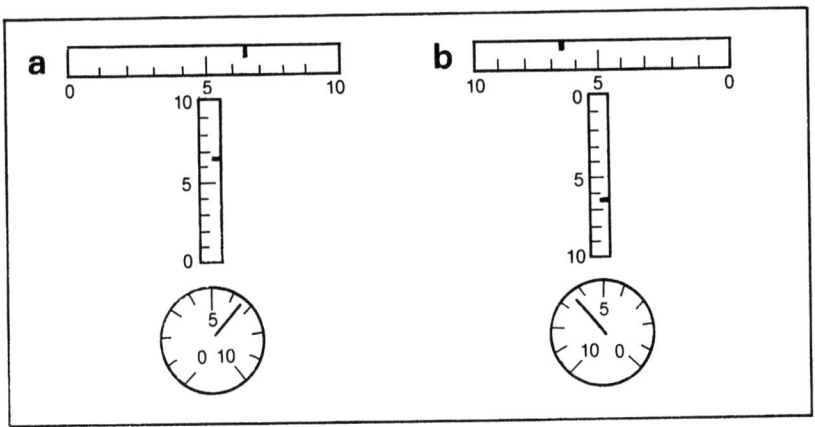

control is in a plane at right-angles to the thing being controlled as shown in figure 28, these unclear relationships should be avoided by better design. Particularly confusing situations arise if there are mixed relationships on the same machine.

More fundamentally there can be no adequate response without the appropriate stimulus, or, to put it another way, the operator needs information to make his decisions and select his actions. If this information is unclear or misleading he will make mistakes. This can arise at any level from inadequate lighting and too much noise so that he cannot see and hear what is happening, to arrays of signals, either visual or auditory or both, which confuse him by sheer complexity or by poor layout and patterning.

The operator has several different capacity limits. There is a limit to the amount of information he can accept at one instant, there is a limit to the rate at which he can accept a stream of information and there is a limit to the amount he can carry in his mind at one time in relation to the current situation. In terms of presented information these limits are all very small, we are talking of around two or three bits, or basic units of information, per glance, five to ten bits per second or a store of thirty bits compared, for example, with communication engineers who sometimes talk in terms of megabits, or millions of bits. Comparisons of hardware and human performance are never simple in numerical terms because we are describing the subjective world of the human operator which may or may not correspond closely to the real world of the engineer. Leaving numbers aside, however, it remains

Figure 28: a. *Ambiguous control/display relationship;* b. *clear control display relationship.*

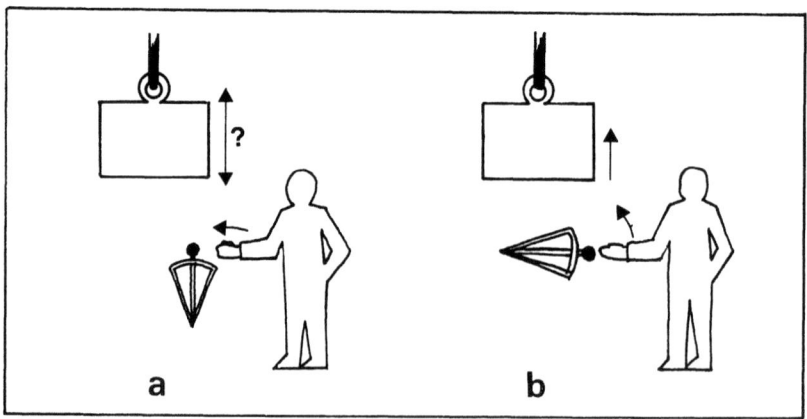

clear that even a skilled operator can easily be overloaded with information. If it emerges from experience that operators are making an unusually large number of errors or are complaining of losing track of what is happening then the situation should be redesigned to reduce the load. In this context contingency planning is crucial because in emergencies the load usually rises suddenly and dramatically which is when effective human performance is most required.

In terms of expected error rates it is again difficult to generalise but it is prudent to assume a fundamental human error rate of about one per cent. At first sight this seems surprisingly high, any machine which failed in some way once in every hundred cycles would be regarded as totally unreliable and yet we know by experience that skilled men are more reliable than machines. The difference is due to the fact that human operators are extraordinarily good at checking their own performance and correcting the errors they have made. This is a very important principle of design. By all means attempt to reduce the initial human error rates, but there are much greater dividends to be had by providing human operators with facilities to detect and correct their own errors.

Related to information theory is decision theory which deals with the situation of ambiguous signals, that is signals with significant associated noise. Detection in these circumstances is often a trade-off in which the operator can reduce the number of missed signals by accepting more signals some of which in fact are false or alternatively if he attempts to reduce the false positives he will inevitably miss more of those to which he should have responded positively, that is omission errors. The terms of this trade-off have to be developed by the skilled operator in the context of the practical consequences of an omission error and a false positive but his behaviour can often be described and predicted using decision theory.

Personality and social theories

Here we are dealing with the emotional man rather than the machine-like human information processor. It was Freud who first highlighted the irrational drives which govern much of our behaviour. He talked in terms of the id and the death instinct, but the terminology is less important than the principle that people can and do sometimes behave aggressively in ways which can lead to damage to themselves and to others. There is extensive evidence about this for behaviour on roads[3] and no doubt these things happen at sea also. The basic point is that not all apparent accidents did in fact happen accidentally.

All kinds of reasons can make individuals deliberately cause damage. It might be failures in their personal lives, which in extreme cases can lead to suicide, or a real or imaginary grievance against the company which can lead to extensive damage to equipment or even danger to other people. Inevitably it is difficult to get at the truth after such events and correspondingly it is not easy to guard against them happening. They are fairly easy to disguise in big systems and the only remedies lie in careful supervision and leadership so that individuals with these tendencies, either permanently or temporarily, can be detected by the ordinary social intercourse between members of the team. For example, if a team leader knows that one of his group is under serious stress because of a crisis in his domestic life, he may for a time find it expedient to avoid putting that man into the most delicate and responsible control situations.

This shades into the general problem of anxiety which also is well documented as a source of human error. Russell Davis has looked in detail at accidents involving road and air crashes from this point of view.[4] He suggests that there are at least three ways in which the behaviour of an experienced skilled operator may fail at a crucial moment so that, for example, he drives past a railway signal at red or he ignores an obvious warning signal on the flight deck. He may have a false hypothesis about what is happening and once having accepted this hypothesis he finds it very difficult to abandon it. If a relaxed but alert operator notes some indication that things are not as they should be he will proceed cautiously and build up his picture of what is happening as the evidence develops; an over-anxious operator, on the other hand, will jump to a conclusion too quickly and he will continue to act on the basis of this conclusion even after considerable evidence has accumulated that what he is doing is wrong. Another phase in which the operator is particularly vulnerable occurs when he has been through a difficult period generating high anxiety but at some point he feels the emergency is over, he thinks he is 'as good as home' and he ceases to attend to what is going on. This may well explain some tanker spills at the end of a long voyage. A second manifestation of the effects of anxiety is in excessive preoccupation with a part of the total picture. Something, perhaps of little consequence, happens which causes him to worry and to concentrate on the event at the expense of much more serious things going wrong to which he does not attend. A third effect of anxiety is the tendency to respond excessively. This has its origin in more primitive man who, when he got into a dangerous situation needed dramatic action in the form of fighting or fleeing. This sudden upsurge of energy still occurs, but it is detrimental to performance in a big system where calm reasoned reactions are called for rather than sudden vicious responses.

The opposite can happen in a social situation where several members of a team are involved in joint action. In order to impress his colleagues with his coolness in emergency, perhaps for the legitimate reason that he is trying to keep them calm, the leader particularly may delay his orders or actions for too long.

Considering social interaction takes us into the very complex field of motivation and the importance of arranging matters so that those involved try just hard enough, not too hard which will lead to the phenomena described above but yet not taking life too easily which leads to carelessness.

The principles of how to maintain the optimal social climate for this purpose have been described by Singleton.[5] Climate seems an appropriate word to describe what must develop because it is not something which can be designed and built but rather it is something which grows and is maintained often by indirect influence. The major influence is the style of the senior people, not only those high in the formal hierarchy but also the skilled and experienced operators from whom newcomers pick up the way things are done in the organisation. Communication is by example as well as by edict. The relevant theories are to do with motivation, leadership, the behaviour of small groups and the behaviour of organisations.[6] Little of this can be applied directly to the business of reducing human error but it is none the less an important facet of the total problem.

Skill and systems theories

It will be clear from what has been said already that human operators are not good performers in emergencies, particularly those involving high speed-stress. Accepting this, there are two complementary strategies, avoid emergencies by anticipating events, which is basically what skill is about, and design for emergencies which is one aspect of what system theory is about.

A skilled human operator is one who, by training and experience, has developed specialised attributes which enable him to match his capacities to his aspirations. His aspirations often involve controlling high power, high speed systems. Yet fundamentally his capacities are limited by a relatively trivial muscular power output and a very complex nervous system which limits his speed of response. He copes with the power problem by utilising external power sources, which incidentally is what the oil industry is all about in its operations and in its objectives. He copes with the speed problem by anticipating events. An unskilled operator accepts a simple stimulus and, after a reaction-time delay, there emerges a response which is more or less appropriate depending on his training and instructions in the form of drills and procedures. This process is inherently slow which means he cannot by this method control either fast-acting systems or massive systems subject to inertial and other lags. To cope with such systems the skilled operator develops strategies which involve modelling what is happening in the outside world. These models may be unconscious and formulated as extensions of his own actions, or they may be simple pictorial ones such as visual images, or they may be symbolic and increasingly abstract ones involving languages and mathematics. By using models the operator escapes from the limitation of functioning in real time, he responds to his model rather than directly to external stimuli and he uses these stimuli together with feedback from his responses to up-date his model. Thus a response is not to an immediately preceding stimulus, but to a whole series of events, some remembered from the immediate past and some which have not yet happened but are predicted by anticipatory techniques involving using his model as a simulator. In this way he reduces the effects of limitations stemming from his basic capacities, including his sluggish reaction time. The effectiveness of the highly skilled operator is no longer dependent on fast reactions to immediate stimuli but rather on the validity of his model or total picture of what is happening. This theory has fundamental implications for training, procedure design, information presentation and for the interpretation of errors. Instead of thinking of the operator as responding to a stimulus we have to think of him as navigating through a multidimensional decision space using an internal map which is maintained by monitoring the pattern of events including the effects of his own actions. He is in phase with real time but not directly latched on to it, he can metaphorically move to and fro into the past and into the future observing, checking and experimenting with what has happened and what might happen. From this theory it becomes clear why untoward incidents can rarely be traced back to single events or stimuli and why certain kinds of fundamental errors occur, for example through using a model that does not sufficiently match reality.

Systems theory is analogous to skill theory in that again we are considering a multi-parameter situation made cohesive by the overriding relationship to a unitary objective. The skilled operator is a particular complex system, but he is also a component of bigger systems such as man-machine systems. For many purposes it is appropriate to begin an analysis of an undesired output such as a flow of polluting oil by looking first at the behaviour of the system from which it comes. This is different from an engineering or hardware based analysis because the human operators are treated as a part of the system, called the personnel sub-system. This has its specialised design problems such as manpower allocation and selection, training and procedure design. There are also design problems in ensuring effective communication between the hardware and the personnel usually called problems of man-machine interfaces.[7] System failures can often, in fact usually, be traced back to failures in these aspects of the total system design. A separate variety of systems theory looking at socio-technical systems uses this approach to describe the effects of technology on the organisations and routines which govern the interactions between people.[8]

METHODS OF ANALYSIS

The preceding description of related theory is a necessary background to the understanding of current methodology and techniques for the description of human errors.

Statistical methods

Collecting and interpreting statistics about errors and accidents is a notoriously difficult business. There are at least three fundamental snags. Firstly, we all like numbers because superficially they seem to provide unambiguous information; but in fact numbers imply categories and determining categories in which to put errors is a complex business. Table 12 shows seven different simple ways of classifying errors and four different types of error classification.[9] Each of these has its value for particular purposes but none is universally applicable. Secondly the source of any evidence is always a great many different individuals with different motivations for being truthful, vague or downright dishonest who have to usually fill in or omit to fill in a variety of different forms.

Table 12: Classifications of error

Kinds of error	Classes of error
Commission: Omission	By cause
Reversible: Irreversible	By effect
Systematic: Random	By responsibility
Detectable: Undetectable	By remedy
Formal: Substantive	
Inputs: Outputs: Decisions	
Recoverable by Machine: Man: Neither	

Table 13: Accidents in offshore installations 1971-75

	Fatal	Serious
Construction	2	6
Drilling	11	61
Production	0	5
Maintenance	1	12
Diving	9	3
Helicopters	0	0
Boats	5	8
Cranes	4	25
Domestic	0	0
Unallocated	0	11
TOTAL	32	131

Source: Development of the oil and gas resources in the UK, Department of Energy.

One must always ask questions such as 'just what is the motivation for that person to be accurate and valid in what is reported' and usually the answer is 'not very satisfactory', similarly 'just what is the justification for assuming the data are comprehensive' and again there are rarely satisfactory grounds for making this assumption. Thirdly, there is no relationship except chance between the grossness of the error in any sense, such as inattention, indiscipline or incompetence and the consequences of the error in terms of damage to people, equipment and the environment.

Most of these problems are minimised when one is dealing with fatal accidents which, for this reason, are a good indicator of the state of the industry in a safety context. Such data for offshore installations and attendant vessels are shown in table 13. For the same period the Health and Safety Commission reported 365 seamen killed on merchant ships registered in the UK. The number of injuries to merchant seamen for 1975 alone was 1,467. A useful comparison with other industries is provided by the risk in terms of fatal accidents per year for each 100,000 employees. For UK registered merchant ships this figure is 116; whilst in 1974 the figure for manufacturing industry is about 4, for agriculture about 12, for the construction industry about 18, for coal mines about 25 and for offshore oil installations about 180 rising to 600 for divers in 1975. (Health and Safety Executive and Department of Energy data). The number of divers employed is only about 500 and this last figure indicates three killed, one fatality more or less would alter the numerical risk very much but nevertheless the total picture is one of an unsafe industry with considerable scope for reducing risks.

Accident investigations

One has to be equally cautious in the interpretation of evidence from inquiries into single accidents or incidents which were unusually dramatic and on a large scale. Who set up the inquiry and why? Is there any way of assessing the particular mix of objectives which usually includes a search for culpability to determine who shall pay what to whom, a whitewashing exercise to mollify the public and a genuine desire to find out what happened so as to avoid a repetition? Were the underlying values legal, economic, scientific, engineering or humanistic and what was the relative emphasis? How many different interests were represented and how strongly did they make and protect their particular case? Since incidents such as spillage from wrecked tankers can take place in any part of the world, there are bound to be considerable national differences in the style of the inquiry. It is essential to establish this context before interpreting a particular report.

There is now some attempt to standardise evidence and statistics by relying on the pollution insurers who are much smaller in number than the shipping companies,[10] but it will take time and experience before this data-gathering yields useful conclusions. There is also some attempt to consider problems of oil-rigs on a world-wide basis by the International Labour Office but, again, it is too early to expect definitive recommendations.[11]

Accident investigations and the supporting governmental and legal infrastructure take a long time to develop a significant effect on and control of an industry. This is a necessary part of the development of a safe and stable industry but it is not sufficient, particularly in the early stages of one which is new or rapidly changing. We need also more direct methods.

Activity analysis

The direct methods involve the study of what workers actually do in practice in the various jobs needed to further the requirements of the industry. These are variously called job analysis, task analysis, error analysis and skills analysis. The underlying procedures are the same, the aim is to describe the activity in detail and the number of ways of gathering evidence is relatively small, these ways are shown in table 14. One can observe and examine the structure and record observations. If there is a strong emphasis on receiving information and decision making, then method-study techniques which concentrate on actions may not be adequate and it will be necessary to use the more elaborate sensori-motor and decision charts. Less directly, one can film what is happening or record it on time-line charts using appropriate kinds of sensing devices such as pressure cells, photo-cells and micro-switches. Communicating with the operator verbally involves interview techniques which vary from simply asking him to talk about what he is doing as he does it, through asking him to discuss critical incidents which he has experienced to the depth interview which might reveal evidence which the operator himself could not or would not reveal spontaneously. Finally there is always a great deal of information and stored records such as log-books, equipment manuals and so on.

Table 14: Job analysis methods and techniques

Method	Technique
Direct Observation	Process-charts, micro-motion charts, flow charts.
Observation and Conference	Sensori-motor charts, decision charts.
Recording	Filming, instrumentation.
Interview	Continuous commentary, critical incident, depth interview.
Records	Data extraction and collation.

A detailed look at the activities of an individual or a team by whatever selection of the above techniques seems most appropriate can be an invaluable starting point for the indentification of potentially unsafe activity. The approach can be very broad or it can be very specific, looking at errors only. An error analysis begins by selecting the appropriate way of classifying errors from the variety of possible ways shown in table 14. It is then a matter of finding the best source of evidence or mixed sources of evidence which will provide the data which can be used to examine what has happened in the past in the hope of reducing the probability that such errors will happen in the future. There are also more elaborate techniques looking, for example, at the risks which operators took and assessing whether or not these were reasonable. The background here is that unlikely events do happen occasionally and it is possible for an accident to happen when everyone involved took perfectly legitimate chances but they were unlucky. The separation of bad luck from incompetence is, of course, a skilled business requiring extensive knowledge of the industry.

It will be clear that there is no standard drill for carrying out any kind of activity analysis. It depends on the kind of job, the seriousness and urgency of the problem, the resources available including money, time and the skill and experience of the analysts.

ERROR REDUCTION PROCEDURES

Human performance

The principle here is to recognise human limitations and accept that if any operator is functioning near a limit he is more likely to make errors. There are many kinds of limits but they can all be avoided by sensitive design of workplaces, tasks and working conditions.

One way of summarising the variety of variables involved is shown in table 15. The operator has physical limitations of size, posture and energy expenditure. He works most effectively when the environmental conditions are kept within fairly narrow bands of temperature, lighting and noise. Adjustments can however be made by varying one of several parameters in relation to each stressor. For example, the effects of high temperature can be mitigated by providing good ventilation, correspondingly in conditions of low temperature the key variable is the amount of air movement from wind and exposure. Vibration is more readily tolerated by the standing operator rather than the sitting operator although, of course, dynamically sprung and damped seats can provide some protection. Vibration and noise are sources of information as well as stress so that insulating the operator completely can be detrimental to his performance. In large-scale systems it is easy to overlook the importance of facilitating access, reach and manipulation and for allowing for variation in the size of operators. Similarly the flexibility and sensitivity of human senses, particularly vision and hearing, can lead to unjustified assumptions, e.g. that one man must look in two directions at once or that he will hear a particular warning in a gale. The human memory is very good for principles, strategies and for monitoring the flow of events but relatively poor for data and other abstract information.

Finally it must be remembered that the operator is subject to cumulative effects; noise, cold, lack of sleep, fatigue etc. all interact with each other usually to the detriment of performance.

Personality and social aspects

The best-known procedures for attempting to make people more careful had their origin in the American space projects, they are called 'zero defect programmes'. These are essentially propaganda exercises aimed at ensuring that workers are aware of the conse-

Table 15: Limitations of human performance

Limitation	Precautions to minimise errors
Physical size	Use data about body size, reach and manipulation dimensions to ensure reasonable fit including some flexibility for different sized operators.
Energy expenditure	Avoid excessive physical work particularly when there are postural limitations. Check that rest periods and shift lengths have been specified with due regard to fatigue.
Environmental conditions	For heat stress provide adequate rest-pauses and ventilation even air-conditioning if the task or stress is critical. For cold stress examine and optimise clothing and length of exposure. Providing protection from wind and wet is more important than avoiding low temperatures. Consider problems of handling tools and materials. Noise and vibration can be fatiguing and can interfere with communication.
Sensory inputs	Noise and vibration are sources of information so that the operator must not be completely insulated from the working scene. There is no excuse for bad lighting. Good lighting incorporates directional aspects, variation and contrast as well as illumination level.
Information inputs	Consider information flow in terms of amount available at any time, rate of presentation and required speed of response. Provide as much anticipatory information as possible. Remember that the operator only needs to know about the past and even the present in order to predict the future.
Information storage and decision making	Consider the coding and structure of data in relation to the required decision. Do not rely on the operator's memory for data, he can confuse things that sound the same even for data seen; new items can interfere with old items, he finds sequences more difficult to recall than content.
Outputs	Design controls so that it is clear what they do. Provide checks so that the operator can confirm that he has done what he wanted to do. Separate selection of action from action controls, preferably with a check between them. Provide him with the maximum flexibility to reverse errors and redirect sequences which are going astray. Check that control dynamics does not lead to instability or excessive skill requirement.

quences of errors on their part. The expertise in this field is to do with keeping the programme running indefinitely at just the right level so that workers do not get bored or antagonistic and yet their attention is not allowed to drift away from these matters. A careful mix of films, demonstrations, posters, statistics, magazines, leaflets and so on is used together with less direct methods such as safety committees and the elections to them and even regular research investigations which require extensive worker participation. It has been noted by some research teams who set out with the specific aim of acquiring evidence that probably their main influence was not in the data and theories they produced, but rather in the increased awareness of safety and error problems which the investigation itself generated.

Similarly the routine procedures for reporting and investigating accidents may well have more influence as an indicator of the importance which senior management attaches to these matters rather than as results leading to conclusions. Training in safety sensitivity can be carried out for various levels of management and it is particularly important for immediate supervisors. How this is done at present is largely a matter of common sense and knowledge of the industry rather than theory.

It should not be forgotten that communication about social aspects of work is not confined to the working situation and is probably more widespread when people are off duty than when they are on duty. Thus the total social life of the workforce, their facilities for interaction and in fact their whole way of life is relevant to style of working including carefulness.

Style of leadership is also important but this is a very complex matter and varies with the traditions of the industry, the kind of work, incentives and disciplinary procedures as well as the personalities of the individuals concerned.

It is, of course, desirable to avoid conflicts between safe working and other objectives such as pay, production or the meeting of deadlines. Again, there is little that can be said in terms of theories or principles except that where there are obvious clashes of interest it is likely that these are consequences of some poor design of organisation and procedures.

Skill and systems aspects

The level of skill of the personnel is of one of the major insurances against system failures. This begins with the design of procedures and here there are always difficult decisions as to how far to rely on skilled individuals to create new solutions to novel problems as they emerge and how far to use contingency planning so that there are drills to cope with most of the things that can go wrong. Resolving these issues is a specialist matter for each industry. On the whole more damage is done to big systems by positively doing the wrong thing than by omitting to do the right thing, so that one key aspect of training is to ensure that individuals do not try to cope with situations which are beyond their particular skill level. A good general rule on the same principle is, 'if in doubt do nothing except await further information'. The provision of adequate information is, of course, crucial for the practice of any skill and, using the model of the skilled man described earlier in this chapter, it is very important to consider the operator's picture of the situation and the ways in which this can be reinforced and developed in phase with the real events. In current jargon, the skilled man in a big system is a supervisor rather than an operator. One way of summarising his function is shown in figure 29. The system is automatic in the sense that it has its own control loops but these are monitored by presenting the appropriate information in displays and controls for the supervisor, he can override the automatic system by manual control and he may also have facilities for trying out actions in simulation through his own local loop before interfering with the real system. In a system of this kind the supervisor can function manually when he wishes to or when his supervisor thinks he should and this provides a way of

Figure 29: *Schematic representation of supervisory control*

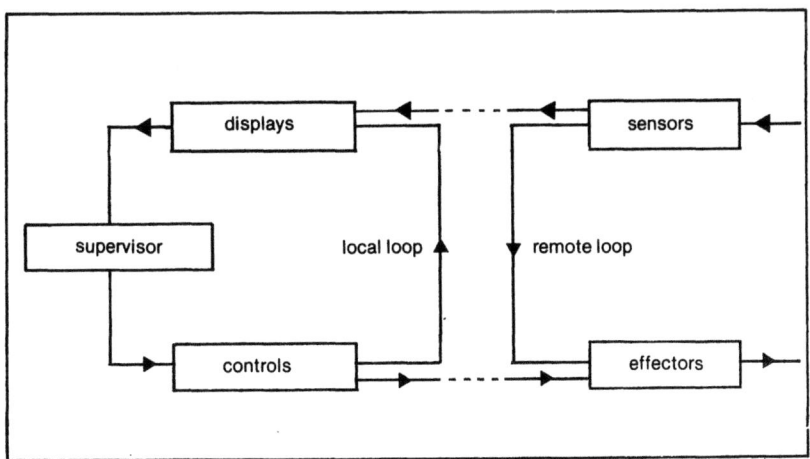

Table 16: Total error reduction strategy

1. Optimal allocation of function between man and machine.
2. Design of procedures again with allocation to man (drills) and machine (software).
3. Selection and training of personnel with consideration of use of otherwise overqualified personnel to increase safety margins.
4. Contingency planning and design or procedure manuals.
5. Design of interfaces and work spaces.
6. Human and hardware based monitoring and manual checking.
7. Consideration of working hours and personnel processes such as grading and promotion.

dealing with the key problem of the maintenance of rarely practised skills. It is easy to invite disaster by expecting an operator to carry out some elaborate task manually under time stress, the requirement for which has arisen because the automatic system has failed.

This approach also deals neatly with the fundamental system design problem which is allocation of function between man and machine. Getting the allocation optimal is a key design feature of a reliable system and supervisory control provides a flexible solution where functions are not rigidly allocated to either man or machine. They are mostly done by machine but the man can get involved in a wide range of them when it is appropriate that he should do so. This enables him to maintain his skills for emergencies and provides him with a greater sense of involvement.

Allocation of function cannot be described entirely in terms of man and machine, procedures can be regarded as a separate set of functions which require a specific allocation and a great deal of design effort. Given that there are three separate areas of function then there are corresponding communication problems between all pairs, but particularly between man and machine. Effective displays and controls are the necessary components of this man-machine interface. The systems approach provides a way of structuring the total error reduction programme as shown in table 16. These are not necessarily carried out in the order shown, the process is usually iterative in that any one decision can interact with and modify others. This comprehensive systems approach to error minimisation is now increasingly practised in high technology industry, the first industry to develop it by experience was civil aviation but it is now spreading into other process industries such as power generation and chemicals and hence to the whale oil industry.

CONCLUSION

This chapter has been deliberately written from a human operator point of view rather than a process point of view partly because there is as yet very little literature specifically devoted to the human factor in oil pollution avoidance problems. It would have been possible to consider separately the human factors aspects of oil-rigs, tankers and refineries and there are now studies going on in all of these areas but there are not yet very many completed achievements which can be referred to. However, it is hoped that the point of view here presented is convincing, that people are people, wherever they happen to work, and that their proneness to error has general features which we understand reasonably well. This understanding can be used in practice to make a considerable impact on reducing system failures and increasing safety.

REFERENCES

1. For a fuller discussion see W.T. Singleton, 'Theoretical approaches to human error', *Ergonomics*, 16.6 (1973), 727-737.
2. A.T. Welford, *Fundamentals of skill* (Methuen, London, 1968)
3. M.H. Parry, *Aggression on the road*, (Tavistock, London, 1978)
4. D. Russell Davies, 'Human errors and transportation accidents', *Ergonomics*, 2 (1958), 24-33.
5. W.T. Singleton, *Human aspects of safety*, Keith Shipton Developments Study no. 8, (Adelaide House, London, 1976)
6. P.B. Warr, *Psychology at work*, (Penguin, Harmondsworth, 1971)
7. W.T. Singleton, *Man-machine systems*, (Penguin, Harmondsworth, 1974)

8. F.E. Emery, *Systems thinking*, (Penguin, Harmondsworth, 1969)
9. W.T. Singleton, 'Techniques for determining the causes of error', *Applied Ergonomics*, 3.3 (1972), 126-131.
10. J. Wardley-Smith, 'Oil spills from tankers' Paper presented at the Marine Ecology and Oil Pollution Conference, Aviemore, (International Tankers Owners Pollution Federation Ltd, London, 1975)
11. International Labour Office, Meeting of experts on safety problems in the construction and operation of offshore drilling installations in the petroleum industry (ILO, Geneva, 1977)
12. W.T. Singleton, 'The ergonomics of information presentation', *Applied Ergonomics*, 2.4 (1971), 213-220.

13

The role of law in the prevention of oil pollution

Professor E.D. Brown, BL, LLM, PhD
Head of Law Department
University of Wales Institute of Science and Technology

As the earlier chapters in this book have shown, oil pollution can and does occur at any stage from exploration to final consumption. The object of this chapter is to complement the description of the technical means of preventing pollution from these various sources, provided in chapters 1-12, with an account of the network of international and municipal (that is, national) laws which have been adopted to prevent oil pollution from these sources.

These laws are many and various, ranging from multilateral conventions of universal relevance and application through regional and even bilateral conventions, reflecting the needs of the area concerned, to national law and regulations, some of which transform the international obligations into municipal law while other supplement the network of international rules. The means of enforcement are also multifarious. The enforcement of international law against foreign ships involves difficult questions of jurisdiction. The enforcement of municipal rules is in some cases a matter of criminal law and of civil law in others.

It is not of course possible in one brief chapter to present an exhaustive analysis of this vast and growing body of rules. The object is rather to be more than superficial but less than exhaustive in presenting a survey which will give the reader an understanding of how the various pieces of the legal jigsaw fit together; to give a bird's eye view of international oil pollution law; to show by reference to selected examples of municipal law and regulations, how treaties are transformed into statutes and regulations; to illustrate the processes of enforcement upon which the efficacy of the law depends.

The most logical sequence would seem to be to follow the oil from exploration all the way to consumption. Accordingly, the second section deals with the exploration and production of oil, including its transportation by pipelines; the third section is concerned with its marine transportation by tankers, while the fourth section refers to the pollution which may occur in various ways after the oil has been delivered ashore.

Scope of 'preventive' oil pollution law

Before turning to the law relating to particular sources of oil pollution, it may be useful to explain the sense in which 'prevention' is perceived for the purpose of this chapter.

The law, both municipal and international, considers particular acts of pollution to be legal 'wrongs'. In municipal law, a somewhat arbitrary distinction is made between civil wrongs and criminal wrongs. Thus, under English law, the owner of an oyster bed which is damaged by oil negligently discharged into the sea would have a civil remedy in tort and could sue for compensation.[1] The wrongdoer (the tortfeasor) might also be guilty, however, of a criminal offence under the Prevention of Oil Pollution Act 1971 and would be liable on conviction to a fine.[2] That the distinction is arbitrary is illustrated by the fact that, in certain circumstances, the criminal courts may make a compensation order against a convicted person for loss or damage resulting from the offence.[3] From the point of view of the protection of the environment, it is clear that the threat of a 'civil' award of damages is no less a deterrent than the threat of the criminal imposition of a fine. It follows that 'preventive' oil pollution law should be regarded as embracing both aspects. Similarly, in international law, it will not suffice to consider only the more obviously 'preventive' conventions such as the International Convention for the Prevention of Pollution of the Sea by Oil 1954-71. The various conventions dealing with civil liability must also be considered.

International and municipal law distinguished

It may also be useful for the non-lawyer if a brief introductory explanation is provided of the chief differences between international law and municipal law and of the relationship between the two systems.

International law is the body of legal rules regulating the relations between international 'persons', that is, entities such as sovereign States which have the legal capacity of having rights and duties under international law. The individual and the company do not possess international personality. The principal sources of international law are international customary law and treaties (or conventions). Evidence of a rule of international customary law — which consists of a general practice accepted as law — is usually sought in the decisions of courts and tribunals and the practice of States. The jurisdiction of international courts and system of compulsory adjudication as in municipal law.

Municipal law, on the other hand, is the law within the State and regulates the relations of individuals (natural persons) and juristic persons such as companies. Although reference will be made to other sources of municipal law (such as the judge-made common law of England) by far the most important or 'laws' or 'decrees' adopted by national legislatures. It is through this source that the international treaty obligations of the State are transformed into rules of municipal law and thus made effective. For example, in the United Kingdom, the Prevention of Oil Pollution Act 1971 is based in part on the International Convention for the Prevention of Pollution of the Sea by Oil 1954, as amended in 1969. Needless to say, the jurisdiction of municipal courts does not depend upon the consent of the parties!

EXPLORATION AND EXPLOITATION OF THE CONTINENTAL SHELF

In both international law and municipal law, the rules which contribute to the prevention of pollution arising from the exploration and exploitation of the continental shelf relate for the most part either to the conduct of offshore operations or to liability for pollution caused by these operations. In this section, the conduct rules will be considered first, followed by the liability rules.

The Conduct Rules

Rules for the conduct of offshore operations which are relevant to pollution prevention are to be found in a number of international conventions and in international customary law.

International conventional law

The 1958 Geneva Conventions
The coastal State exercises over the continental shelf sovereign rights for the purpose of exploring it and exploiting its natural resources.[4] Though described as 'sovereign', these exclusive rights are enjoyed only for the exploration and exploitation of continental shelf resources and have to be exercised in accordance with other limiting rules. For example, the coastal State must not impede the laying or maintenance by other States of submarine cables or pipelines, subject to its right 'to take reasonable measures for the exploration of the continental shelf and the exploitation of its natural resources'.[5] More generally, the continental shelf activities of the coastal State must not result in any unjustifiable interference with navigation, fishing or the conservation of the living resources of the sea, nor result in any interference with fundamental oceanographic or other scientific research carried out with the intention of open publication.[6]

As regards prevention of oil pollution, there are a number of relevant provisions in the Geneva Conventions, though it must be said that they are exceedingly vague.

Firstly, every State is required to draw up regulations to prevent pollution of the seas by the discharge of oil from pipelines or resulting from the exploration and exploitation of the seabed and its subsoil. Such regulations are to be drafted 'taking account of existing treaty provisions on the subject'.[7]

Secondly, the coastal State is entitled to construct, operate and maintain 'installations and other devices' necessary for exploration and exploitation of continental shelf resources, to establish around them the 500-metre safety zones which ships of all nationalities must respect, and to take in those zones measures necessary for their protection.[8] The installations and devices are declared to be 'under the jurisdiction of the coastal State'.[9] Due notice must be given of the construction of such installations and permanent means for giving warning of their presence must be maintained. Moreover, abandoned or disused installations have to be entirely removed.[10] Installations must not be established where interference may be caused by them or their safety zones to the use of recognised sea lanes essential to international navigation.[11] Finally, the coastal State is required to undertake in the safety zones all appropriate measures for the protection of the living resources of the sea from all harmful agents.[12]

Regional Conventions

On an even more general level, provision is made for the prevention of pollution from offshore operations in the two regional conventions for the Baltic and the Mediterranean.[13] Thus in the Baltic Convention, each State Party accepts an obligation to take all appropriate measures to prevent pollution from exploration or exploitation of its part of the seabed and its subsoil or from any 'associated activities thereon'. It must also ensure that adequate equipment is at hand to start an immediate abatement of pollution in the area.[14] Similar provision is made for the Mediterranean.[15]

The International Convention for the Prevention of Pollution from Ships, 1973 and the 1978 Protocol

The 1973 Convention[16] relates to oil pollution caused by 'installations and devices' operating on the continental shelf to only a limited extent. The release of harmful substances directly arising from the exploration, exploitation and associated offshore processing of seabed mineral resources is specifically excluded from its scope.[17] Nonetheless, special requirements are laid down for the prevention of *operational* discharges of oil. Fixed and floating rigs, when engaged in the exploration, exploitation and associated offshore processing of seabed mineral resources, must comply with the rules applicable to ships of 400 tons and above other than oil tankers.[18] They are, therefore, prohibited from discharging oil or an oily mixture into the sea except when the installation is more than 12 miles from land and is not within a specified special area.[19] The oil content of the effluent must be less than 100 parts per million and the installation must be equipped, so far as is practicable, with an oil discharge monitoring and control system, oily water separating equipment and tanks for oil residues (sludge).[20]

Because of the complexity of the Convention and the costs which would be involved in bringing it into operation, the conviction was widespread until very recently that the Convention was increasingly unlikely ever to enter into force. The prospects may, however, have been changed by the adoption in February 1978 of a Protocol to the Convention.[21] *Inter alia*, the Protocol will enable States to ratify the parent Convention by accepting only Annex 1 — the Regulations for the Prevention of Pollution by Oil — and not also, as was originally provided, Annex 2 — the Regulations for the Control of Pollution by Noxious Liquid Substances in Bulk. This modification, together with the pressures for action generated by a series of casualties off the coasts of the United States in 1976 and 1977 — intensified by the *Amoco Cadiz* disaster in March 1978 — may well lead to the reasonably early entry into force of the Convention.

Other conventions

As has been noted, however, the 1973 Convention is of only limited relevance to the prevention of oil pollution from continental shelf operations, and unfortunately, there is no comparable international convention which prescribes detailed regulations for these operations. The feasibility of establishing uniform standards for construction and use of offshore installations and possibly incorporating them in a treaty has been investigated by a Working Group which met in March 1973.[22] There was apparently wide agreement on the standards set by the Working Group but so far the work has not led to the adoption of any international agreement.

Pending the conclusion of such an international convention, the regulation of the construction and operation of offshore installations and devices has been left to each State and examples of such national provision are considered below.

To a minor extent, the national legislation has been co-ordinated by bilateral treaties. For example, the Anglo-Norwegian Agreement of 22 May 1973 relating to the oil pipeline from the Ekofisk field to the United Kingdom has provisions concerning the routeing of the pipelines, the desirability of harmonising national safety requirements and removal of disused pipelines.[23] Similarly, the Anglo-Norwegian Frigg Agreement of 10 May 1976 (dealing, however, only with the transmission of natural gas from the Frigg Field Reservoir) contains pollution safeguards.[24]

International customary law

Some of the conventional limitations upon the exercise by the coastal State of its sovereign rights to explore and exploit the natural resources of the continental shelf have been examined above. It must now be added that this negative right to exclude others from the continental shelf (for these specified purposes) also has as its positive counterpart the duty under international customary law to protect the interests of other States and their nationals which may be injured in the exercise of these exclusive rights. For it is a fundamental rule of international customary law that 'A State owes at all times a duty to protect other States against injurious acts by individuals from within its jurisdiction.'[25]

Or, as it has been put, 'under the principles of international law no State has the right to use or permit the use of its territory in such a manner as to cause injury by fumes in or to the territory of another or the property or persons therein, when the case is of serious consequence and the injury is established by clear and convincing evidence'.[26] If the continental shelf is not the territory of the coastal State, it is at least 'within its jurisdiction' in relation to exploration and exploitation of natural resources.[27]

What, however, is the extent of the coastal State's duty? There would seem to be no evidence for the contention that, under international customary law, the coastal State is strictly liable (that is, liable even in the absence of negligence or want of reasonable care) for such oil pollution, on the ground that the exploration of the continental shelf and the exploitation of its natural resources (or any of them) constitute an ultra-hazardous activity.[28] The duty is rather to exercise reasonable care in protecting the interests of foreign States and their nationals. More specifically, the coastal State would be required to provide regulations for the exploration and exploitation of the continental shelf, the reasonable adequacy of which would be determined by reference to 'good oilfield practice'[29] and, in relation to navigational hazards, by the standards of good seamanship. The standard of reasonable care would also require adequate State supervision of compliance with such regulations.

Clearly, the limitations placed on the coastal State's sovereign rights by these rules of international customary law are on a level of generality comparable with that of the limiting rules in the Geneva Convention, refered to above.

Summing up the international rules
Summing up then, it has been seen that the exercise of the sovereign rights enjoyed by the coastal State for the exploitation of the continental shelf are limited by a number of rules of treaty and customary law which, directly or indirectly, contribute to the prevention of pollution from offshore oil operations. These rules require:
1. no unjustifiable interference with navigation, fishing or the conservation of the living resources of the sea;
2. the drawing up of regulations to prevent pollution from pipelines or exploration and exploitation of the sea-bed and subsoil;

3. measures for the protection of safety zones around installations or devices on the continental shelf;
4. notice of construction of installations, means of warning of their presence and removal of disused or abandoned installations;
5. measures to protect living resources in safety zones;
6. a prohibition of operational discharges;
7. compliance on installations with the standard of care observed in good oilfield practice.

In addition, the bilateral Ekofisk and Frigg Agreements make provision for routeing of pipelines and construction and safety standards.

So far, it has been left very largely to municipal law to develop these vague rules of conventional and customary law by transforming them into precise and detailed rules and regulations and to make provision for their enforcement. The object of the following section is to illustrate this process.

Municipal law
It is hardly feasible to review within the confines of this chapter the provision which the various national legal systems have made for the regulation of continental shelf operations. It is possible, however, to present a useful model of how these problems can be tackled on the national level by concentrating primarily on the laws and regulations of the United Kingdom and complementing this account with comparisons drawn from the practice of other States.[30]

The law of the United Kingdom provides for the prevention of oil pollution from offshore operations through a complex network of statutes and delegated legislation. An outline is given below of the principal statutes, regulations and orders:

The Prevention of Oil Pollution Act 1971
This Act consolidated a number of earlier statutory provisions and, in particular, replaced that part of the Continental Shelf Act 1964 which dealt with oil pollution from pipelines and offshore installations.[31] The Act makes it an offence to discharge oil or oily water as a result of exploration or exploitation of seabed resources[32] but in practice only a 'best practicable' standard is demanded.[33] Discharges from pipelines are also prohibited.[34] Though strict liability applies to deliberate discharges, it is a

defence, where the discharge consists of an escape of oil, to prove that reasonable care has been exercised.[35] The maximum fine is £50,000 on summary conviction and an unlimited fine on conviction on indictment. Officers of operating companies can also be proceeded against in certain circumstances.[36]

The Continental Shelf Act 1964
Though in part superseded by the 1971 Act, some provisions of this Act are still relevant to oil pollution. By Orders made under the Act, 500-metre safety zones around installations have been designated 'prohibited areas' to which access is limited.[37] Provision is made for the application in the offshore area of the civil and criminal law.[38] Finally, the Act enables the Department of Trade to control the siting of installations to ensure that there should be no obstruction or danger to navigation.[39]

The Petroleum (Production) Regulations 1976
These Regulations[40] provide the principal means by which the Government controls the exploration and exploitation of the continental shelf. They embody Model Clauses on safety and related matter for inclusion in exploration and exploitation licences.[41] Under such Clauses, the consent of the Secretary of State is required for starting, abandoning or re-starting a production well.[42] The Institute of Petroleum has prepared a Code of Safe Practice for the industry.

The Mineral Workings (Offshore Installations) Act 1971
This Act empowers the Secretary of State for Energy to make safety regulations for offshore installations and imposes certain responsibilities upon installation managers. Reflecting the need for continual review and revision of safety regulations, the Act is drafted in general terms and its objectives largely achieved through subordinate legislation.[43] As a result of regulations made under the Act, all offshore installations, fixed or mobile (even rigs classified as ships), must be registered — thus enabling the Department of Trade to maintain a proper record.[44] More importantly from the point of view of pollution, all installations must have a Certificate of Fitness, indicating compliance with the design and construction standards laid down.[45]

Under the Act, the installation manager has far-reaching powers and responsibilities for safety, health or welfare and maintenance of order and discipline. He must not allow any operation likely to endanger the seaworthiness or stability of the installation.[46]

The Petroleum and Submarine Pipelines Act 1975
Part III of the Act is of particular relevance to the pollution question.[47]

Under the Act, no person may construct a pipeline in the territorial sea or the designated areas of the continental shelf or use such pipeline without the written consent of the Secretary of State for Energy.[48] The Minister can thus control the numbers and routeing of pipelines. In issuing an authorisation, the Secretary of State may impose such terms as he thinks appropriate including, *inter alia,* terms as to 'the steps to be taken for the purpose of ensuring that funds are available to discharge any liability for damage attributable to the release or escape of anything from the pipeline'.[49]

The Secretary of State has far-reaching powers to make regulations to secure the proper construction and safe operation of pipelines, prevention of damage to pipelines and the safety, health and welfare of persons engaged on pipeline works.[50]

Contravention of the terms of a pipeline authorisation renders the holder liable to termination of the authorisation and it is a criminal offence, punishable by fine, to construct or use a pipeline without authorisation or to contravene any provision of certain notices which may be served on the pipeline owner.

Comparison with Norwegian law
According to a recent detailed study,[51] the Norwegian approach to control of operational discharges from offshore installations is indistinguishable from that of the United Kingdom. As in the United Kingdom, the standard is dictated by technology and Norwegian policy is to require the use of the best available technology.

As regards accidental discharges, the Norwegian rules are contained in exceedingly detailed Decrees of 1975 and 1976,[52] dealing with exploration and production, and in related Regulations. The main differences between the two systems are that: a) construction standards are set in the United Kingdom by classification societies whereas in Norway the Petroleum Directorate sets them itself; and b) unlike the United Kingdom, Norway does not make extensive use of Model Clauses in licences.

The Liability Rules

International customary law
Reference was made above[53] to the duty of the coastal State under international customary law to exercise reasonable care in protecting foreign States and their nationals from oil pollution damage as a result of the exploration or exploitation of its continental shelf. As was seen, the rule is of a very general nature and the recovery of compensation for a breach of the rule would never be easy, being dependent *inter alia* on either the tortfeasor State's free acceptance of its responsibility or its consent to accept some form of compulsory third-party settlement such as arbitration or adjudication by the International Court of Justice. On the level of international law, the claim would be that of the 'victim' State rather than of the local authorities, firms and individuals more directly injured by the pollution. It would be the responsibility of the State to co-ordinate and press these claims upon the tortfeasor State.

Such vague rules are of course quite unsatisfactory and it is the object of the 1976 Convention, referred to below,[54] to provide precise, detailed rules on such matters as standard of liability ('fault' or 'strict'), limitation of liability and jurisdiction. Under such a conventional regime, it then becomes practicable for the actual victims of oil pollution damage to sue in the national courts of the States Parties to the Convention. As Parties, they are under an obligation to ensure that their courts have jurisdiction in such cases and are empowered to award compensation in accordance with the standards set by the Convention.

International conventional law
Under the Helsinki Convention on the Protection of the Marine Environment of the Baltic Sea Area (1974),[55] the Parties undertake jointly to develop and accept rules on responsibility for damage resulting from acts or omissions in contravention of the Convention (including those causing pollution from offshore operations). *Inter alia,* such rules would deal with limits of responsibility, criteria and procedures for the determination of liability and available remedies. A similar undertaking is contained in the Barcelona Convention for the Protection of the Mediterranean Sea against Pollution (1976).[56] The desirability of uniformity of standards on this question is self-evident and it is all the more important that their development should take place

under these treaties since, as will be seen below, the 1976 Convention on Civil Liability — the only treaty so far to deal with this matter in detail — failed to provide uniform limits of liability.

Convention on Civil Liability for Oil Pollution Damage resulting from Exploration for and Exploitation of Seabed Mineral Resources, 1976[57]
Final agreement on the text of this Convention proved to be very difficult to secure and the end product is less than satisfactory, particularly in relation to the limits of liability.

Liability of operator. The operator of an installation at the time of an incident is liable for any pollution damage arising from the incident[58] and pollution damage is defined to mean 'loss or damage outside the installation caused by contamination resulting from the escape or discharge of oil from the installation and includes the cost of preventive measures and further loss or damage outside the installation caused by preventive measures'.[59] 'Installation' is so defined as to cover all fixed or mobile installations and platforms, storage installations and pipelines provided they, or a substantial part of them, are situated seaward of the low-water line.[60]

No claim outside the Convention may be made against the operator[61] but it would seem possible in appropriate cases to sue subcontractors such as manufacturers of defective equipment and thus avoid the limits of liability laid down in the Convention.

Strict Liability. Liability is strict. The operator will thus be liable, without need to prove fault or negligence on his part, unless: a) the damage resulted from an act of war, hostilities, civil war, insurrection, or a natural phenomenon of an exceptional, inevitable and irresistible character (acts of terrorism or sabotage do not fall within this exception); or b) the incident occurred more than five years after a well was abandoned under the authority, and in compliance with the requirements, of the coastal State; or c) the victim intentionally or negligently contributed to the damage.[62]

Limited Liability. Provided the pollution damage is not caused by the deliberate act or omission of the operator, with actual knowledge that pollution damage would result, the operator may limit his liability for each installation and each incident to the amount specified in the Convention.[63] Unfortunately, agreement could not be reached on an appropriate figure and, as a result, the Convention will not in fact achieve one of its objectives which was to ensure an equal degree of protection to victims of pollution in all the Convention countries, irrespective of the location of the incident or of the damage.

Liability may be limited to approximately $35 million until 1 May 1982 and to approximately $47 million thereafter.[64] States Parties have the option, however, to provide for unlimited liability or a higher limit of liability. The ceiling of liability may therefore vary from country to country — clearly an undesirable position.

To cover his liability, the operator is bound to carry insurance or provide other financial security of not less than approximately $26 million until 1 May 1982 and approximately $41 million thereafter. However, the operator may be exempted from having to carry insurance cover for damage caused by an act of sabotage or terrorism.

There is no automatic adjustment of the limits of liability and insurance to reflect inflation but a Committee is established with power to recommend changes. Since parties are free not to accept such recommendations, this procedure provides another potential source of different limits.

Actions for compensation may be brought in the courts of any State Party where pollution damage has been suffered or in the courts of the coastal State within whose offshore area the offending installation is situated.[65] An operator wishing to limit his liability under the Convention is required to constitute a fund equal to the total amount of his liability with one such court and that court then has exclusive competence to determine all matters relating to the apportionment and distribution of the fund.[66]

Despite its shortcomings in relation to the limit of liability, this Convention does embody a reasonable balance between the interests of potential victims of oil pollution damage and the interests of the operating oil companies and it is to be hoped that it will shortly enter into force and form the basis of the necessary municipal legislation of the States Parties.

The Offshore Pollution Liability Agreement (OPOL)
In the meantime, victims of oil pollution damage may be able to obtain compensation under the Offshore Pollution Liability Agreement (OPOL),[67] a contract concluded by a number of oil companies on 4 September 1974 which entered into force on

1 May 1975. The main features of the Agreement, as amended on 12 September 1975 and 23 March 1976, may be summarised as follows:

1. Participating companies accept strict liability for damage and clean-up costs arising from their operations to a maximum of $25 million per incident.
2. Exceptions to liability are similar to those provided for in the 1976 Convention referred to above but also include acts or omissions done with intent to cause damage by a third person. This would of course exclude liability for acts of sabotage or terrorism.
3. The Agreement is open to operators of offshore facilities situated within the offshore areas of the following States: Denmark, France, Ireland, the Netherlands, Norway and the United Kingdom.
4. Of the $25 million, $12.5 million is earmarked for damage claims and the other half for clean-up costs.
5. Parties to OPOL must provide proof of financial responsibility by way of insurance or other security for the full $25 million or show that they are capable of self insurance. If, nevertheless, any operator is unable to meet claims under the Agreement, the balance up to the $25 million ceiling would be met on a shared basis by the other OPOL members.
6. Provision is made for arbitration of disputed claims in accordance with the rules of the International Chamber of Commerce.
7. Claims may be made by any person for pollution damage and by a 'public authority' for clean-up costs. Public authorities include both the national government and local authorities.

Municipal law
Pending the entry into force of the 1976 Civil Liability Convention and the enactment of implementing legislation, there is no statute in the United Kingdom which makes provision for the recovery of compensation for oil pollution damage arising from offshore oil operations. It may, however, be possible to recover damages by suing under the general law of torts. An action may be founded on negligence or, possibly, on nuisance.

As regards negligence, it would be necessary to show that the discharge or escape of oil was the result of failure to observe 'good oilfield practice' and that the resultant oil pollution damage was foreseeable.

There is some doubt whether a private nuisance may emanate from the open sea (as distinct from the usual case of neighbouring property)[68] but a better foundation would in many circumstances be provided by the concept of public nuisance, that is, anything 'which obstructs or causes inconvenience or damage to the public in the exercise of rights common to all Her Majesty's subjects'.[69] The commission of a public nuisance is a criminal offence but may also found an action in tort. The Attorney-General may seek an injunction to terminate or prevent a recurrence of the nuisance and any individual suffering greater damage or inconvenience from the oil than the generality of the public would have an action to recover damages.[70]

The position in Norway, the other most important North Sea oil-exploiting State, is not dissimilar.[71] Pending the entry into force of legislation implementing the 1976 Convention, liability is governed by the general law of torts. Norwegian tort law is in general based on fault liability unless 'hazardous' activities are involved and it would appear that the standard of strict liability does apply to offshore oil operators. Strict liability is specifically imposed on licensees. Under the present law, there is no ceiling on liability and the licensee is vicariously liable jointly and severally with the tortfeasor and the tortfeasor's employer.

MARINE TRANSPORTATION BY TANKERS

The rules which contribute to the reduction of oil pollution damage caused by ships fall into one of three categories dealing respectively with the conduct of tankers, the powers of control enjoyed by coastal States, and liability for oil pollution damage.

The Conduct Rules

General international law

Prior to the dawning of the new environmental era in the early 1950s, relatively little concern was expressed about the established procedure whereby the ballast carried in the cargo tanks of oil tankers was discharged into the sea prior to taking on a new load.

Once the public's interest in the environment was aroused, the questions put to the lawyers were: Is the discharge of oil into the sea in this way lawful? Does general international law contain rules prohibiting the use of the seas for such refuse disposal? Is freedom to wash oil tanks into the sea one of the freedoms of the high seas recognised by international customary law? Though the position is not entirely clear, it would seem that, under international customary law, the right of the user of the high seas is not an absolute right but only a right of reasonable user.[72] This view is confirmed by the Geneva Convention on the High Seas (1958), which is 'generally declaratory of established principles of international law'.[73] Article 2 provides that the freedom of the high seas must be 'exercised by all States with reasonable regard to the interests of other States in their exercise of the freedom of the high seas'. It would seem to follow that the unreasonable discharge of noxious materials into the high seas is prohibited. But international customary law offers no further guidance as to what substances are noxious and what would amount to an unreasonable discharge. Nor is Article 24 of the Geneva Convention of much greater assistance, providing only that 'Every State shall draw up regulations to prevent pollution of the seas by the discharge of oil from ships ... taking account of the existing treaty provisions on the subject'. However, it provided a starting point for the more specialised treaties referred to below.

Brief mention must also be made of the Stockholm Declaration on the Human Environment.[74] Principle 7 'states the common conviction'[75] that 'States shall take all possible steps to prevent pollution of the seas by substances that are liable to create hazards to human health, to harm living resources and marine life, to damage amenities or to interfere with other legitimate uses of the sea'.[76]

Specialised international conventional law
The conduct rules contained in the network of treaty provisions and recommended practices which the international community has developed to prevent oil pollution from ships relate to, first, the intentional discharge of oil, and, secondly, to accidental spillages.

The intentional discharge of oil

THE INTERNATIONAL CONVENTION FOR THE PREVENTION OF POLLUTION OF THE SEA BY OIL 1954-1971.
The basic instrument is the International Convention for the Prevention of Pollution of the Sea by Oil 1954-1971. The convention was amended in 1962, 1969 and, in minor respects, in 1971.[77] The 1969 amendments entered into force on 20 January 1978.

The basic formula. The Convention distinguishes between tankers and ships other than tankers. The details of the permitted discharges from tankers and the problems that arise have already been considered in chapter 6 at pp119-120, discharge of oil or of any oily mixture from machinery spaces in other ships is discussed at p.127.

Facilities for disposal of residues. The prohibition of discharges of oil into the sea is only a realistic proposition on the assumption that adequate facilities are provided at ports (including repair ports) and oil loading terminals for the disposal of oil residues. The Convention accordingly requires States Parties to 'take all appropriate steps to promote the provision of such facilities'.

Enforcement. To facilitate enforcement of the Convention, every ship to which the Convention applies and which is either a tanker or a ship using oil fuel is required to carry an oil record book, and specified entries must be made in it.[78] The oil record book may be inspected by the competent authorities of any Contracting State — but only while the ship is within a port in that State and subject to the authorities acting as expeditiously as possible and not delaying the ship. A true copy may be taken of any entry in the oil record book and may be admitted as evidence in any subsequent judicial proceedings.

It is also provided that any contravention of the ban on discharges of oil or of the obligation to maintain an oil record book shall be a punishable offence under the law of the flag State. Penalties to be imposed for unlawful discharges outside the territorial sea must be adequate in severity to discourage such discharges and not less than the penalties for the same infringements within the territorial sea.

Although only the flag State may bring proceedings in respect of an infringement of the Convention, any Contracting Government may furnish it with particulars in writing of evidence that any provision has been contravened, wherever the alleged offence took place. A Government so informed is required to investigate the matter, and if satisfied that sufficient evidence exists, to institute proceedings as soon as possible and report to the other Government and IMCO on the action taken.[79]

Policing the ban. It is one thing to establish a ban on specified discharges of oil; it is quite another to ensure that offenders will be detected, identified and prosecuted. The 1969 amendments to the Convention have introduced marginal improvements in this respect but the traditional policing model remains unchanged, with only the flag State having the right in international law to interfere with ships on the high seas or to institute proceedings against the owner or master. Proposals have been made to give the 'port-State' the power to prosecute for offences committed on the high seas but they have not yet been adopted.[80]

THE INTERNATIONAL CONVENTION FOR THE PREVENTION OF POLLUTION FROM SHIPS 1973 AND THE 1978 PROTOCOL

The detailed rules on the prevention of oil pollution which were intended to replace those of the 1954-71 Convention are contained in Annex 1 of the 1973 Convention — the 'Regulations for the Prevention of Pollution by Oil'.[81] The main innovations are also discussed in chapter 6, but mention must also be made of the provision made for 'special areas' of enforcement.

Special areas. Five areas — the Mediterranean, the Black Sea, the Baltic, the Red Sea and the 'Gulfs' area — have been designated as special areas in which only discharges of 'clean ballast' are allowed.[82]

Enforcement. Violation of the Convention must be forbidden and punishments provided for under the municipal law of the Parties. To ensure that the Convention is enforced, provision is made for three types of jurisdiction.[83] First, the 'flag State' of the vessel may prosecute for offences wherever they occur. Secondly, the coastal State may prosecute for offences committed within its jurisdiction. Unfortunately, the geographical scope of this competence will remain obscure until the U.N. Conference on the Law of the Sea completes its work because Article 9 (3) of the Convention says simply that the term 'jurisdiction' 'shall be construed in the light of international law in force at the time of application or interpretation of the prevent convention' — a formula necessitated by failure to agree upon where the limits lay.[84] Thirdly, the Convention provides for 'port State' jurisdiction but only to a very limited extent.

Entry into force. As has been seen, the poor prospects for the early entry into force of the Convention have been somewhat improved by the adoption of the Protocol of 1978,[85] which will enable States to ratify the Convention while accepting only Annex I (Oil Pollution Regulations). Considerable obstacles have still to be overcome, however, especially in making available improved oily-water separating and monitoring equipment and shore-based reception facilities. Deficiencies in these respects still prevent States from ratifying the Convention.

ICS Pollution Prevention Code (Oil Tankers). Pending its entry into force, the 1973 Convention is not without effect in practice since some eighty companies have undertaken to abide by a Code designed by the International Chamber of Shipping to permit as many as practicable of the provisions of the Convention to be brought into operation on a voluntary basis (see also chapter 6 p.120).

The accidental discharge of oil

As the *Amoco Cadiz* catastrophe has recently reminded the world, the control of intentional discharges of oil into the sea will do nothing to reduce the damage done by oil spills resulting from marine casualties such as collision and stranding. Reference must also be made therefore to the rules on safety at sea.

Following the *Torrey Canyon* disaster in 1967, a comprehensive work programme was agreed upon in the IMCO Council. It included under the heading of preventive measures to be considered: sealanes; shore guidance; speed restrictions; navigational equipment; officer and crew training; use of automatic pilots; construction and design of tankers, and identification and charting of hazards.

Much progress has been made since 1967 on these topics and work continues. In 1967 the main conventional rules for the prevention of accidents were contained in the International Convention for the Safety of Life at Sea (SOLAS 1960) and the Collision Regulations 1960 appended to it.[86] Since that time, the Convention on the International Regulations for Preventing

Collisions at Sea 1972[87] has entered into force (on 15 July 1977) and SOLAS 1974 has been adopted, though it has not yet entered into force. As regards crew training, an International Convention on Standards of Training, Certification and Watchkeeping for Seafarers was adopted at an IMCO Conference in July 1978.

SOLAS 1960 (the version still in force) contains chapters relating to *inter alia* the following aspects of safety: a) ship construction, including subdivision and stability, machinery and electrical installations and fire protection; b) radiotelegraphy and radio telephony; and c) safety of navigation. SOLAS 1974 includes, *inter alia,* new regulations on fire protection for tankers and provides a more speedy procedure for the adoption of further amendments. A Protocol to the Convention, adopted in 1978, introduces further advances in relation to inert gas systems, steering gear and radar and collision avoidance aids.

The entry into force of the 1972 Collision Regulations in 1977 was a major step forward in maritime safety. Revised in the light of recent technological advances, the Regulations recognise the increased use of radar and contain provisions to ensure that very large vessels are not hampered by other ships in confined areas. Most importantly, however, the new Regulations provide for mandatory observation of the previously advisory traffic separation schemes. Over 100 schemes have been adopted by IMCO,[88] almost exclusively in areas of congested or converging traffic. Prior to 1977, a number of major maritime States had voluntarily introduced legislation requiring vessels flying their flag to observe such routeing schemes and could therefore institute proceedings against violators on receipt of evidence from other States. Under the 1972 Convention, however, Parties are now legally bound to adopt the necessary municipal legislation.[89]

Municipal law
The Prevention of Oil Pollution Act 1971
The obligations of the United Kingdom under the International Convention for the Prevention of Pollution of the Sea by Oil 1954, as amended in 1969, are implemented in the Prevention of Oil Pollution Act 1971.

It is an offence under the Act for the owners or master of a ship registered in the United Kingdom to discharge oil into the sea outside British territorial waters, except in compliance with the *en route*, rate-of-discharge and distance-from-land rules laid down in the 1969 amendments to the 1954 Convention.[90] It is also an offence to discharge oil of any kind into the territorial waters of the United Kingdom or those parts of its internal waters which are navigable by sea-going ships.[91] Either the owner or master or, it has been held,[92] both may be charged with such offences. It is, however, a defence to prove (i) that the discharge was for the purpose of securing the safety of the vessel, or for preventing damage to any vessel or cargo or for saving life, unless the court is satisfied that such discharge was not necessary or was not a reasonable step to take; or (ii) that the oil or mixture escaped as a result of damage to the vessel, and that as soon as practicable after the damage occurred all reasonable steps were taken to prevent, stop or reduce the escape; or (iii) that the oil or mixture escaped by reason of leakage, that neither the leakage nor any delay in discovering it was due to any want of reasonable care, and that as soon as practicable after the escape was discovered all reasonable steps were taken for stopping or reducing it.[93] Offenders are liable to fines of up to £50,000 on summary conviction and of unlimited amount on conviction on indictment.

Regulations have been adopted prescribing the installation of certain equipment in ships to prevent oil pollution.[94]

Provision is made for the maintenance and inspection of an oil record book [95] and for the supply by harbour authorities of facilities for the disposal of oily residues.[96]

The Merchant Shipping Act 1974
Part II of this Act provides for the implementation of the 1971 amendments to the 1954 Convention on tank arrangements and limitation of tank size and empowers the Secretary of State to make 'oil tanker construction rules' in implementation of the 1971 amendments or of any other convention signed for the United Kingdom — even though it is not in force for the United Kingdom.[97] The Act also makes provision for the survey and certification of tankers to ensure compliance with the construction rules[98] and places restrictions upon uncertified tankers as regards sailing from or entering British ports.[99] Failure to comply with a prohibition on entry renders the owner and the master liable to fines.

Collision regulations
The 1972 Collision Regulations have been implemented by an Order made under the Merchant Shipping Act 1894[100] which makes mandatory for British ships traffic separation schemes adopted by IMCO. Wilful infringement of the regulations is a criminal offence but a more effective incentive is provided by the

fact that, in civil proceedings following a collision, a departure from the regulations will normally provide evidence of negligence.

Merchant Shipping Bill 1978

At the time of writing (December 1978), the Merchant Shipping Bill is passing through Parliament. If enacted, it will enable the United Kingdom to ratify the International Convention for the Prevention of Pollution from Ships 1973 and other Conventions dealing with oil pollution from ships.[101]

Powers of Control of Coastal States

International law

Under international customary law, the scope of the coastal State's powers of interference with a *Torrey Canyon* type casualty — a vessel causing or threatening serious coastal pollution following collision or stranding — is uncertain, though the adoption of necessary preventive measures can probably be justified under the rules on necessity.[102]

It was to clarify these powers that the Brussels Intervention Convention was adopted in 1969.[103] Under the Convention, Parties 'may take such measures on the high seas as may be necessary to prevent, mitigate or eliminate grave and imminent danger to their coastline or related interests from pollution or threat of pollution of the sea by oil, following upon a maritime casualty or acts related to such a casualty which may reasonably be expected to result in major harmful consequences'.[104] The related interests include commercial fishing and tourist interests, the health of the coastal population and the conservation of living marine resources and of wild life.[105] The measures taken must be 'proportionate to the damage actual or threatened' and 'reasonably necessary'.[106] Measures going beyond what is reasonably necessary and causing damage to others render the coastal State liable to pay compensation.[107]

Municipal law

British law

The United Kingdom adopted implementing legislation — the Oil in Navigable Waters Act 1971 — even before the Convention entered into force and the rules are now contained in the consolidating statute, the Prevention of Oil Pollution Act 1971 and in the Oil in Navigable Waters (Shipping Casualties) Order 1971.[108]

Under the Act, the Secretary of State has power of direction over the owner, master and salvor of the ship to move or not to move the ship; to unload or not to unload cargo; and to take or not to take specified salvage measures. If such powers are inadequate, the Secretary may proceed to 'any action of any kind whatsoever', including destruction of the vessel. This applies to ships registered in the United Kingdom wherever the casualty occurs and to ships registered abroad where it occurs in the British territorial sea; the Order extends these powers to ships registered abroad where the accident occurs outside the British territorial sea.

French law
Ratification of the Brussels Intervention Convention was authorised by a French Law of 1971.[109] As the *Amoco Cadiz* incident has shown, however, the powers of control recognised by the Convention are of little use unless the coastal authorities are informed promptly of the existence of a situation which may lead to large scale pollution damage and this will be required for ships entering the English Channel as from 1 January 1979.[110]

Regional Co-operation in Dealing with Oil Spills

The Bonn Agreement for Co-operation in Dealing with Pollution of the North Sea by Oil (1969).[111] was very much a response by the eight North Sea Contracting Parties to the *Torrey Canyon* affair. The basic objective of the Agreement is protection against oil pollution which presents a grave and imminent danger to the coast or related interests of one or more of the Parties and 'active co-operation' is the chosen means.

Such co-operation is of two kinds — preparatory and organisational co-operation and co-operation following a casualty. Under the first head, provision is made for exchange of information on competent authorities and means to deal with pollution and eight zones of responsibility are established.

Prescribed co-operation following a casualty is of three kinds: provision is made for the speedy gathering of information; for monitoring and distribution of information by the zonal authority; and for assistance in dealing with the spill.[112]

A similar agreement was concluded among the Nordic countries in 1977.[113]

The Liability Rules

International law
The Brussels Conference which produced the Intervention Convention also adopted the International Convention on Civil Liability for Oil Pollution Damage 1969,[114] the purpose of which is to provide uniform international rules and procedures for determining questions of liability and providing adequate compensation to persons who suffer damage caused by the escape or discharge of oil from ships.[115]

Under the Convention, liability is placed upon the owner of the ship.[116] Liability is strict (no need to show fault or negligence) and the onus rests on the owner to prove that any of the exceptions should operate.[117] The exceptions relate to such matters as acts of war, natural phenomena of an exceptional, inevitable and irresistible character, wrongful acts and negligence of the victim and failure by the authorities to maintain lights or other navigational aids.[118] Except where he has been guilty of fault, the owner may limit his liability per incident to approximately £86 per ton of his ships gross tonnage, with a ceiling of approximately £9,054,123[119] and ships must be covered by insurance or other financial security for the owner's total liability for one incident.[120] To ensure observance of the Convention, ships are required to carry certificates, confirming the existence of such insurance. States Parties must not permit ships flying their flags to trade without insurance and must ensure that vessels entering or leaving their harbours and other installations in the territorial sea do carry insurance.[121] The Convention applies to damage caused in the territory of a contracting State, including its territorial sea, even when the party liable is a non-contracting State or its national. It does not apply to damage caused beyond the territorial sea, except that brought about by preventive measures designed to minimise damage inside the territorial sea.[122]

Actions may be brought in the courts of the State or States in which damage has been suffered and an owner wishing to limit his liability must establish a fund with one such court which will then have sole competence to determine all matters relating to the apportionment and distribution of the fund.[123]

Recognising that the insurance cover provided for under the Convention would offer inadequate protection to potential victims, the Parties adopted a supplementary instrument, the International Convention on the Establishment of an International Fund for Compensation for Oil Pollution Damage, 1971.[124] The

1971 Convention established a fund financed by levies on persons who import or receive oil in contracting States. It is designed to provide supplementary compensation for pollution damage to the extent that the protection afforded by the 1969 Convention is inadequate and to relieve shipowners of the additional financial burden imposed on them by that Convention (additional, that is, to their liability under a 1957 Limitation of Liability Convention). The amount of compensation payable per incident is limited so that the aggregate paid under the 1969 Convention and the Fund should not exceed approximately $35 million, though the Fund Assembly is empowered to raise that figure to a maximum of $70 million.[125]

Municipal law
The provisions of the 1969 Civil Liability Convention are implemented in the United Kingdom by the Merchant Shipping (Oil Pollution) Act 1971. The Act provides that:

> where, as a result of any occurence taking place while a ship is carrying a cargo of persistent oil in bulk, any persistent oil carried by the ship (whether as part of the cargo or otherwise) is discharged or escapes from the ship, the owner of the ship shall be liable, except as otherwise provided by this Act, —
>
> (a) for any damage caused in the area of the United Kingdom by contamination resulting from the discharge or escape; and
>
> (b) for the cost of any measures reasonably taken after discharge or escape for the purpose of preventing or reducing any such damage in the area of the United Kingdom; and
>
> (c) for any damage caused in the area of the United Kingdom by any measures so taken.[126]

The exceptions are those provided for in the Convention and referred to above.

'Contamination' is not defined and it is uncertain whether economic loss suffered by seaside hoteliers would fall within the scope of the damage covered by the Act, though it is reasonably certain that holidaymakers would have no remedy for spoiled holidays.

Following the Convention, the Act restricts claims against the owner to those made under the Act and exempts the owner's servants and 'any person performing salvage operations' with his agreement from liability[127]. Actions may still be brought under the common law against persons other than the owner, but their liability would not be strict.[128]

Ships entering or leaving a British port or offshore terminal in the territorial sea are required to carry the certificate of insurance prescribed by the Convention and any entry or departure in violation of the Act is an offence rendering the owner *and/or master* liable to fines of up to £35,000 on summary conviction and of unlimited amount on conviction on indictment.[129] Ships attempting to leave a United Kingdom port without a valid certificate may be detained and this applies even to ships registered in States which are not parties to the Convention.

The 1971 Fund Convention is implemented in the United Kingdom by Part I of the Merchant Shipping Act 1974, though the sections providing for compensation have not yet been brought into operation.[130] One weakness of the Act is that the Fund is not liable if the claimant cannot prove that the damage was caused by a ship identified by him.[131]

Industry schemes

Compensation may also be available under the voluntary schemes introduced by the oil tanker industry.

The Tanker Owners Voluntary Agreement concerning Liability for Oil Pollution (TOVALOP) came into force on 18 September 1969[132] but has been amended several times, most recently in June 1978. TOVALOP provides that a Participating Tanker Owner will compensate persons, including governments and local authorities, who sustain pollution damage as a result of the escape or discharge of oil. Persons taking preventive measures to mitigate such damage are also covered. The Agreement further provides for compensation for costs incurred by any person in taking measures to remove the *threat* of a discharge of oil, even if no discharge occurs. Liability is strict and subject to the same limitation and defences as in the Civil Liability Convention, 1969.[133] No liability arises under TOVALOP when there is a right of recovery under the Civil Liability Convention. In case of dispute, liability may be enforced through arbitration under the Rules of the International Chamber of Commerce.

TOVALOP is supplemented by CRISTAL — the Contract Regarding an Interim Supplement to Tanker Liability for Oil Pollution Damage, signed in 1971 and most recently amended in June 1978. Although originally intended to provide an interim supplement to the Civil Liability Convention and TOVALOP until the Fund Convention came into force,[134] CRISTAL still supplements these and other sources by providing compensation for pollution damages not otherwise recoverable. CRISTAL is funded by contributions from its cargo-owner Parties and liability is conditional upon the oil being owned by a Party and the tanker in question being enrolled in TOVALOP. The circumstances must be such that the Civil Liability Convention imposes liability on the tanker, or would have done if it had been applicable. The ceiling on liability under CRISTAL is the same as under the Fund Convention1[135] and is subject to deduction of sums recovered from other sources of compensation.

OIL POLLUTION ON LAND

It is not possible within the confines of this chapter to give a detailed account of the law dealing with discharges of oil on land or into inland waterways, resulting from the transport of oil by road or rail, its storage and distribution, and its consumption in vehicles, factories and homes. Every country has its own, often very different, network of laws, regulations and institutions designed to curb such discharges.[136] It must suffice to refer briefly to a few of the attempts which are being made on a regional level to deal with the problem and to indicate by reference to United Kingdom practice the variety of municipal laws and regulations required to prevent or minimise oil pollution from these various sources.

International Law — Regional Conventions

The Paris Convention for the Prevention of Marine Pollution from Land-based Sources, 1974,[137] adopted by ten western European States,[138] is a regional instrument through which the Parties undertake to combat marine pollution from land-based sources by adopting individually and jointly the measures prescribed in the Convention.[139] 'Pollution from land-based sources' is defined to include pollution (i) through watercourses, (ii) from the coast, including introduction through underwater or other pipelines, and (iii) from man-made structures placed under the jurisdiction of a Party within the limits of the Convention area.[140]

More particularly, the Parties undertake to *eliminate* pollution by substances listed in an annexed 'black list' of substances and to *limit strictly* pollution by substances in a 'grey list'. The black list includes 'persistent oils and hydrocarbons of petroleum origin' and the grey list 'non-persistent oils and hydrocarbons of petroleum origin'.[141]

As regards enforcement, each Party undertakes to ensure compliance with the Convention and to take in its territory appropriate measures to prevent and punish conduct in contravention of the provisions of the Convention.[142]

The United Nations Environment Programme (UNEP) has set itself the task of having regional Action Plans adopted by 1982 for seven seas: the Mediterranean, the Gulf, the Caribbean, the East Asia Sea, the Gulf of Guinea, the Red Sea, and in the Pacific. It is likely that the Action Plans will include conventions, similar to the Paris Convention, dealing with land-based sources of pollution.[143]

The role played by the European Economic Community (EEC) deserves special mention. Though the EEC is of course a regional organisation, it differs from most other such organisations in that its organs enjoy supranational powers. It is sometimes described as a 'functional federation'. Thus, in relation to environmental questions, the EEC Council is empowered to issue 'Directives', that is, instruments which are binding upon member States as to their objectives while leaving choice as to the means of attaining those objectives. It is clear that an increasingly important contribution to the prevention of pollution will be made by the EEC through its programme of Action on the Environment.[144]

Municipal law

Storage tanks

The type of regulations required to ensure that pollution should not be caused by discharges from storage tanks at the port, the refinery, the filling station, the factory and the home has been described in Chapter 10. The technical problems being broadly similar wherever oil is stored, the draftsman of new regulations will find adaptable models in the legislation to which reference has been made.[145]

Industrial oil wastes

A major source of oil pollution on land is the discharge of oily wastes into sewers and water courses from refineries, petrochemical works and factories. Provision is made to control such discharges in both EEC Directives and in municipal legislation.

The EEC Council has issued a Directive on the disposal of waste oils[146] and, under a proposed new Directive [147] on the Protection of Goundwater against Pollution, the discharge of mineral oils and hydrocarbons of petroleum origin directly into the groundwater zone will be prohibited and indirect discharges through the ground will be subject to authorisation. The proposed Directive would require member States to legislate within two years and to ensure that the regulations were in force within three years.

Perhaps the most effective means of preventing pollution by particular noxious materials is to make their discharge an offence but to provide for the discharge of trade effluents with the consent of a licensing authority and subject to the conditions it may impose. This is basically the system operated by the water authorities in England and Wales, established by the Water Act 1973.[148] Controls may also be imposed at an earlier stage when planning permission is being sought for a proposed development of land. Thus, in exercising its powers under the Town and Country Planning Act 1971, the local planning authority normally consults the water authority if the development is likely to affect inland waters.

The United Kingdom and her EEC partners have been unable to agree on a common approach to the control of discharges. The EEC Commission and the other Member States advocate the adoption of a system whereby uniform discharge standards are laid down, thus controlling pollution at the point of discharge. The United Kingdom, on the other hand, believes that the better approach is to establish water criteria standards appropriate to the circumstances of the areas concerned, and to vary the conditions to which permitted discharges are subject so as to ensure the maintenance of such standards. This approach is maintained in Part II of the Control of Pollution Act 1974 which is to be brought into effect in 1979.[149] Pending the entry into force of Part II of the 1974 Act, the principal statutory controls over the pollution of inland surface waters are to be found in the Salmon and Freshwater Fisheries Act 1923, the Public Health Act 1936 and the Rivers (Prevention of Pollution) Acts 1951-61. Under the latter Acts, it is an offence to cause or knowingly permit to enter a stream any poisonous, noxious or polluting matter, or any matter which tends to impede the flow of water so as to be likely to lead to a substantial aggravation of pollution. The Acts provide a flexible instrument of control, however, in that the authority may grant consent to applicants wishing to discharge trade effluent.[150]

As regards underground waters, it is an offence under the Water Resources Act 1963 to discharge 'by means of a well, borehole or pipe' any trade effluent, or poisonous noxious or polluting matter into any underground stream. Here, too, the water authority may grant consents.[151]

Brief mention should also be made of powers enjoyed by statutory water undertakers to make by-laws under the Water Act 1945 to prevent the contamination of water supplied by them.[152]

The Clean Rivers (Estuaries and Tidal Waters) Act 1960, embodies controls similar to those for non-tidal waters. Discharges are thus controlled by the water authority, using its consent procedure.[153]

Transport and distribution of oil
The principal concern of laws and regulations dealing with the transport of oil by road and rail is to ensure safety but of course safe transport also prevents pollution. A model for road transport is provided by the United Kingdom's Petroleum Spirit (Consolidation) Act 1928 and the Orders made under it.[154] The regulations relate to such matters as the types of containers used, the precautions to be taken and the warning signs and labels to be attached. Similar rules are established by British Rail regulations for transport by rail.

Emissions from motor vehicles
In Europe, uniformity of standards on pollution from vehicle exhausts is gradually being introduced through EEC Directives on vehicle exhaust gases and the sulphur content of liquid fuels.[155] In the United Kingdom, the law is laid down principally in the Road Traffic Acts 1960 and 1972 and in regulations made thereunder, particularly the Motor Vehicles (Construction and Use) Regulations 1973.[156]

CONCLUSION

If progress in the prevention and control of oil pollution could be gauged by reference to the number and complexity of laws adopted to deal with it, the above survey would surely show that considerable advances had been made in the past few years. That the reality is less impressive than the appearance is due largely to the fact, clearly evidenced in the table appended to this chapter, that many of the conventions still lack the necessary number of ratifications to bring them into effect. Moreover, by the time they do come into force, they tend to offer somewhat dated solutions.

Nevertheless, even measured in terms of conventions in force, much progress has been made, and of course the introduction in municipal law of the substance of the conventions does not necessarily await their entry into force. Moreover, on every front work is proceeding. At the time of writing (December 1978), the Seventh Session of the Third United Nations Conference on the Law of the Sea (UNCLOS III) has ended and new rules on tankers have been adopted, following the *Amoco Cadiz* disaster. If, after further sessions of the Conference, a new convention on the law of the sea is adopted, it will probably also include provision for 'port State jurisdiction' and increased pollution jurisdiction for the coastal State in a 200-mile exclusive economic zone.[157] The *Amoco Cadiz* affair has also prompted Governments to re-examine the rules on the conduct and liability of tankers in IMCO and in the EEC.[158] On the regional front, UNEP continues to act as the catalyst in creating further strands in the network of oil pollution agreements. Finally, though less publicised, national legislatures and administrations continue to add their contributions to the legal jigsaw.

It is no longer realistic — if ever it was — to expect that international discharges of oil will be eliminated by the end of the decade;[159] the hope must now be that progress towards this goal will continue and will not rely for stimulus upon more disasters like the *Amoco Cadiz*.

REFERENCES

1 *Foster v. Warblington U.D.C.* [1906] 1K.B. 648.
2 See further below,
3 Powers of Criminal Courts Act 1973, s.35.
4 Under the Geneva Convention on the Continental Shelf, 1958, Article 2(1) and under international customary law.
5 See Geneva Convention on the High Seas, 1958, Article 2 and Geneva Convention on the Continental Shelf, 1958, Article 4.
6 Geneva Convention on the Continental Shelf, 1958, Article 5(1).
7 Geneva Convention on the High Seas, 1958, Article 24.

8 Geneva Convention on the Continental Shelf, 1958, Article 5(2).
9 Article 5(3).
10 Article 5(5).
11 Article 5(6).
12 Article 5(7).
13 The Helsinki Convention on the Protection of Marine Environment of the Baltic Sea Area, 22 March 1974, *International Legal Materials* (ILM), Vol.13, 1974, p.546 ; and the Barcelona Convention for the Protection of the Mediterranean Sea against Pollution, 16 February 1976, ILM, Vol.15.1976, p.290.
14 Helsinki Convention, Article 10.
15 Barcelona Convention, Article 7.
16 Misc. No.26 (1974), Cmnd. 5748; ILM, Vol.12, 1973, p.1319.
17 By the definition of 'discharge' in Article 2(3).
18 Annex 1, Regulation 21.
19 Regulation 9.
20 Under Regulation 9(1)(b)(iii), it would appear that to be so entitled to discharge oil, the installation would, additionally, have to be 'proceeding en route' but, being in clear conflict with both the letter and evident intention of Regulation 21, this is probably a drafting error. See also D. W. Abecassis, *Marine Oil Pollution*, University of Cambridge Department of Land Economy, Occasional Paper No.6, 1976, at pp.52–53.
21 The Protocol of 1978 relating to the International Convention for the Prevention of Pollution from Ships 1973 was accepted at the IMCO International Conference on Tanker Safety and Pollution Prevention (IMCO Document TSPP/CONF/11, 16 February 1978 and TSPP/CONF/11/Add.1, 7 March 1978).
22 Set up by a Conference on Safety and Pollution Safeguards in the Development of North West European Offshore Mineral Resources. See further V. Fitzmaurice, *A Critical Assesment of Pollution Control Laws regulating the Development of Petroleum Resources in the United Kingdom and Norwegian Sectors of the North Sea*, (Ph.D. thesis, University of Edinburgh 1977, unpublished), pp.141–143.
23 U.K. Treaty Series No.101 (1973), Cmnd. 5423; ILM, Vol.13, 1974, p.26, Articles 7, 8 and 10.
24 Cmnd. 6491 (1976), Articles 17 and 23.
25 C. Eagleton, *Responsibility of States in International Law*, 1928, p.90, cited with approval by the Arbitral Tribunal in the *Trail Smelter Arbitration* (U.S. — Canada, 1938; 1941), 3 R.I.A.A., p.1905, at p.1963.
26 *Trail Smelter Arbitration, loc. cit.* in note 25, at p.1965. See also the *Corfu Channel Case (Merits), I.C.J. Reports, 1949*, p.4, at p.22, where reference is made to 'every State's obligation not to allow knowingly its territory to be used for acts contrary to the rights of other States'.
27 See also Stockholm Declaration of the U.N. Conference on the Human Environment (16 June 1972,ILM, Vol.11, 1972, p.1416). Principle 21 provides that 'States have, in accordance with the Charter of the United Nations and the principles of international law, the sovereign right to exploit their own resources pursuant to their own environmental policies, and the responsibility to ensure that activities within their jurisdiction or control do not cause damage to the environment of other States or of areas beyond the limits of national jurisdiction."
The Declaration is not a treaty and has no binding force of itself. As a vehicle for the expression of the *opinio juris* of States, however, it may contribute to the formation of rules of customary law.
28 See further on the principle of fault in international law, B. Cheng, *General Principles of Law as Applied by International Courts and Tribunals*, 1953, Chap.8, especially at

pp.225–226 and G. Schwarzenberger, on International Law, Vol.1, 1957, Chap.35. See also, the analogous question of the basis of liability for pollution of international rivers, A. Lester 'Pollution', in A.H. Garretson, R.D. Hayton, and C.J. Olmstead (eds.), *The Law of International Drainage Basins*, 1967, For arguments in favour of the introduction of absolute liability, see L. F. E. Goldie, 'Liability for Damage and the Progressive Development of International Law', *International and Comparative Law Quarterly*, Vol.14, 1965 and Goldie, 'International Principles of Responsibility for Pollution', *Columbia Journal of Transnational Law*, Vol.9,1970.
29 On 'good oilfield practice', see text below, following note 40.
30 On the practice of other European States, see the 13-volume series *The Law and Practice relating to Pollution Control in the Member States of the European Communities* published by Graham and Trotman Ltd.
31 S.5.
32 S.2.
33 Fitzmaurice, *op.cit.* in note 22, at p.532.
34 S.3.
35 S.6.
36 S.19(8).
37 Under s.2. The Continental Shelf (Protection of Installations) Order 1978, S.I. 1978/260, which came into operation on 31 March 1978, revoked all previous Protection of Installations Orders and consolidated the substance of them in one instrument.
38 S.3. and the following Orders: The Continental Shelf (Jurisdiction) Order, S.I. 1968/892; the Continental Shelf (Jurisdiction) (Amendment) Orders, S.I. 1971/721; S.I. 1974/1490; S.I. 1975/1708; S.I. 1976/1517.
39 S.5(4), extending to designated areas of the continental shelf the provisions of Part II of the Coast Protection Act 1949.
40 S.I. 1976/1129.
41 Schedules Five and Seven.
42 Schedule 5, Model Clause 17(1)–(2).
43 The following Orders have been issued under the Act: a Commencement Order, S.I. 1972/644 and the following Offshore Installations Regulations:- Registration, S.I. 1972/702; Managers, S.I. 1972/703; Logbooks and Registration of Death, S.I. 1972/1542; Inspectors and Casualties, S.I. 1973/1842; Construction and Survey, S.I. 1974/289; Public Inquiries, S.I. 1974/338; Diving Operations, S.I. 1974/1229; Operational Safety, Health and Welfare, S.I. 1976/1019; Emergency Procedures, S.I. 1976/1542; Life-saving Appliances, S.I. 1977/486.
44 Offshore Installations (Registration) Regulations 1972.
45 Offshore Installations (Construction and Survey) Regulations 1974.
46 S.5(4).
47 Part II having been replaced by the 1976 Petroleum (Production) Regulations.
48 S.20 refers only to pipelines the construction of which began before 1 January 1976 – the date of entry into force of this Act.
49 S.21(3)(e).
50 S.26.
51 Fitzmaurice, op.cit. in note 22 at p.532.
52 Royal Decree of 3 October 1975 Relating to Safe Practice etc. in Exploration and Drilling for Submarine Petroleum Resources and Royal Decree of 9 July 1976 Relating to Safe Practice for the Production etc. of Submarine Petroleum Resources. Discussed in Fitzmaurice, Chap.8.
53 In text at notes 25 and 26.

54 In text at note 57.
55 ILM, Vol.13, 1974, p.552, Article 17.
56 ILM, Vol.15, p.290, Article 12.
57 Misc. No.8 (1977), Cmnd. 6791; ILM, Vol.16, 1977, p.1451. The Convention, adopted by an Intergovernmental Conference on 17 December 1976, was open for signature from 1 May 1977 to 30 April 1978 (and for accession thereafter) by the following nine States: Belgium, Denmark, Germany, France, Ireland and Netherlands, Norway, Sweden, and the United Kingdom. Under Article 18, States Parties may unanimously agree to invite other States bordering the North Sea, the Baltic Sea or the Atlantic north of 36°N to accede to the Convention. It might possibly be extended, therefore, to Canada, Finland, German Democratic Republic, Iceland, Poland, Portugal, Soviet Union, Spain and the United States. The Convention will enter into force on the ninetieth day following ratification or accession by four States.
58 Article 3.
59 Article 1 (6).
60 Article 1 (2).
61 Article 4 (1).
62 Article 3.
63 Article 6.
64 Article 6 expresses these limits by reference to the Special Drawing Rights (SDR) of the International Monetary Fund (IMF). It has been customary in such Conventions to express the limits of liability by reference to the 'Poincaré' franc, based on gold. Since, however, gold no longer provides a basis for expressing uniform amounts in different countries, this Convention — and as will be seen, other oil pollution liability conventions — have adopted as a new unit of account the SDR of the IMF. At the time of the adoption of the Convention, 1 SDR was worth approximately $1.17.
65 Article 11.
67 Articles 6(5) and 11(3).
67 Text in R. Churchill et al, (eds.), *New Directions in the Law of the Sea*, New York, Oceana Publications Inc., Vol. VI, 1977, p.507.
68 See further G.W. Keeton, 'The Lessons of the Torrey Canyon", *Current Legal Problems*, Vol.21, 1968, pp.94–112, at p.100.
69 Stephen's *Digest of Criminal Law*, 5th ed., 1894, p.140.
70 See observations of Denning L.J. in *Esso Petroleum Co. Ltd.* v. *Southport Corporation* [1954] 2 Q.B. 182 at 197, cited in Keeton, *loc. cit.* in note 68, at pp.100–101.
71 This account is based on Fitzmaurice, *op. cit.* in note 22, pp.494–496.
72 See further E.D. Brown, *The Legal Regime of Hydrospace*, London, Stevens & Sons, 1971, pp.127–128.
73 Preamble.
74 See note 27 above for citation and on status of the Declaration.
75 Preamble.
76 See also the related Recommendation 86(b) (*ibid.*, p.1454) under which Governments are recommended to ensure *inter alia* that the provisions of existing treaties are complied with by ships flying their flags.
77 For a composite text of the Convention, as amended in 1962 and 1969, see ILM, Vol.9, 1970, p.1.
78 Article IX.
79 Article X.
80 See text below, following note 84 and note 157.

81 The Convention is in ILM, Vol.12, 1973, p.1319 and Annex I is at p.1335.
82 Annex I, Regulation 10.
83 Articles 4—6.
84 See further E.D. Brown, in *Current Legal Problems*, Vol.28, 1975, at pp.208—209.
85 See text above, at note 21.
86 U.K. Treaty Series No.65 (1965), Cmnd. 2812 and No.23 (1966), Cmnd. 2956.
87 Misc. No.28 (1973), Cmnd. 5471.
88 IMCO, *Ships Routeing*, 3rd ed., 1973 and *Supplement 1975*.
89 This is so because, under Article 1 of the 1972 Convention, Parties undertake to give effect to the Rules annexed to the Convention. Rule 10 consists of a list of rules to be observed by ships navigating in or through IMCO traffic separation schemes and Rule 1(a) provides that "These rules shall apply to all vessels upon the high seas and in all waters connected therewith navigable by seagoing vessels".
90 S.1. The Exception is provided for in the Oil in Navigable Waters (Exceptions) Regulations 1972, S.I. 1972/1928.
91 S.2. Dishcarges into the territorial sea may also be offences under Part II, s.31 of the Control of Pollution Act 1974 once Part II is brought into operation. The Government intends to bring Part II into force before the end of 1979 (*The Times*, 14 April 1978).
92 *Federal Steam Navigation Co. Ltd.* v. *Department of Trade and Industry*, [1972] 2 All E.R. 97; [1974] 1 W.L.R. 505, H.L.
93 S.5.
94 The following Regulations have been adopted under S.4: Oil in Nivigable Waters (Ships Equipment) (No.1) Regulations 1956, S.I. 1956/1423 and Oil in Navigable Waters (Ships' Equipment) Regulations 1957, S.I. 1957/1424, which continue to apply under s.33(2) of the Act.
95 Under s.17 and the Oil in Navigable Waters (Records) Regulations 1972, S.I. 1972/1929, maintained in force by s.33(2) of the Act.
96 S.9.
97 Ss.10 and 11, which were brought into force on 1 November 1974 by the Merchant Shipping Act 1974 (Commencement No.1) Order 1974, S.I. 1974/1792. No such rules have been made so far.
98 S.11(3) but the necessary rules have not been issued yet.
99 Ss.12 and 13 which have not been brought into effect.
100 Collision Regulations and Distress Signals Order 1977, S.I. 1977/982, as amended by S.I. 1977/1301.
101 H.C. Bill 1978-79 [13] Merchant Shipping Bill. Under clause 20, 21 and 39, the Government will be enabled to ratify the International Convention for the Prevention of Pollution from ships, 1973 and its 1978 Protocol; the 1978 Protocol to SOLAS 1974; and the 1976 Protocols to the International Convention on Civil Liability for Oil Pollution Damage 1969 and the International Convention on the Establishment of an International Fund for Compensation for Oil Pollution Damage 1971.
102 E.D. Brown, *op. cit.* in note 72, at pp.139—146.
103 International Convention relating to Intervention on the High Seas in Cases of Oil Pollution Casualties, 29 November 1969, U.K. Treaty Series No.77 (1975), Cmnd. 6056; ILM, Vol.IX (1970), p.25.
104 Article I(1).
105 Article II(4).
106 Article V(1) and (2).
107 Article VI.
108 S.I. 1971/1736.

109 Law No. 71–1002 of 16 December 1971. See further C.A. Colliard, *The Law and Practice Relating to Pollution Control in France*, London, Graham and Trotman Ltd., 1976, p.101.
110 Under an agreement reached at a meeting of the Anglo-French Safety of Navigation Group on 5 September 1978, all ships of more than 1600 tons carrying oil through the Channel will be required to report to monitoring stations on entering the separation zones off Dover, Ushant and the Casquets (*The Times*, London, 6 September 1978).
111 U.K. Treaty Series No.78 (1969), Cmnd. 4205; ILM, Vol.9, 1970, p.359.
112 For a more extended analysis, see Brown, *op. cit.* in note 72, at pp.158–162.
113 Agreement between Denmark, Finland, Norway and Sweden concerning Co-operation in Taking Measures against Pollution of the Sea by Oil, 1971 (S.H. Lay *et al*, eds., *New Directions in the Law of the Sea*, Vol.II, 1973, pp.637–640).
114 International Convention on Civil Liability for Oil Pollution Damage, 29 November 1969, U.K. Treaty Series No.106 (1975), Cmnd. 6183: ILM, Vol.IX (1970), p.45.
115 Preamble.
116 Article III(1).
117 Article III(2)–(3).
118 Article III(2).
119 Article V(1) and (2). The limits of liability established by Article V(1) were 2,000 gold francs per ton, with a ceiling of 210 million gold francs. Under the Merchant Shipping (Sterling Equivalents) (Various Enactments) Order 1978, S.I. 1978/54, the sterling equivalents are £86.23 and £9,054,122.94. Under the Protocol to the 1969 Convention, adopted on 19 November 1976 (Misc. No.26 (1977), Cmnd. 7028; ILM, Vol.16 (1977), p.617) the unit of account was changed from the Poincaré franc to the SDR of the IMF, though non-parties to the IMF have the option to retain the gold franc. See also note 64 above.
120 Article VII(1).
121 Article VII(2)–(4), (10) and (11).
122 Article II. See further E.D. Brown, *op. cit.* in note 72 at pp.168–171.
123 Articles V(3) and IX(1) and (3).
124 Misc. No.26(1972), Cmnd. 5061; ILM, Vol.11,1972, p.284; entered into force October 1978.
125 The figures laid down in Article 4 of the Fund Convention were 450 and 900 million gold francs. These figures were converted into 30 and 60 million SDR's by the Protocol of 19 November 1976 (ILM, Vol.16, 1977, p.621 – not yet in force). The sterling equivalents are £19,404,000 and £38,808,000. See also note 119 above.
126 S.1(1). The Act is now fully in force (S.I. 1971/1423 and S.I. 1975/867).
127 S.3, following Article III(4) of the Convention.
128 See text above, around note 68.
129 S.10(2).
130 Ss.2 and 4–8.
131 S.4(7)(b), reflecting the provisions of Article 4(2)(b) of the Convention.
132 The original agreement is in ILM, Vol.8, 1969, p.497; amendments are published by the International Tanker Owners Pollution Federation Ltd. which administers the scheme.
133 See text above, following note 114.
134 The Fund Convention entered into force in October 1978. The original CRISTAL agreement is in ILM, Vol.10, 1971, p.137. Amendments and published by the administering body, the Oil Companies Institure for Marine Pollution Compensation Ltd.
135 See text above, at note 25.
136 See further the 13-volumes series cited in note 30 above, especially J. McLoughlin, *The Law and Practice Relating to Pollution Control in the Member States of the European Communities: A Comparative Survey*, 1976.

137 Misc. No.1 (1975), Cmnd. 5803; ILM, Vol.13, 1974, p.352. The Convention is not yet in force.
138 Denmark, France, Germany, Iceland, Luxembourg, Netherlands, Norway, Spain and the United Kingdom.
139 Article 1(2).
140 Article 3(c).
141 Article 4(1) and Annex A, Parts I and II.
142 Article 12.
143 The Mediterranean Action Plan provides a model. See Final Act of Conference on the Protection of the Mediterranean Sea (1976), ILM, Vol.15, 1976, p.285. Agreement on a complementary Protocol on Land-Based Sources was not reached at the Monte Carlo conference in January 1978.
144 See further the following section of the annual General Report of the European Communities:- Fifth Report 1971, pp.242−245; Sixth Report 1972, pp.198−202, Seventh Report 1973, pp.235−241; Eighth Report 1974, pp.135−139; Ninth Report 1975, pp.136−139; Tenth Report 1976, pp.152−158. See also the following EEC publications:- "The Protection of the Environment; an Urgent Problem for Industrial Societies", *Bulletin of the European Communities*, No.4, 1971, pp.53−58; "A Community Programme for the Environment", Supplement to *Bull. of E.C.*, No.5, 1972; "Environment Action Programme", Supplement to *Bull. of E.C.*, No.3, 1973; "Definition of a Community Environment Policy", *Bull. E.C.*, No.7/8, 1974, pp.11− 14; *State of the Environment. First Report*, 1978.
145 See references appended to Chapter 10.
146 EEC Council Directive No.75/439/EEC (CJ No.L.194, 25.7.75).
147 COM (78) 3, published on 24 January 1978.
148 For a detailed account, see J. McLoughlin, *The Law and Practice Relating to Pollution Control in the United Kingdom*, Graham & Trotman, 1976, especially Chap.3.
149 As announced in a House of Commons written reply by the Minister of State at the Department of the Environment (*The Times*, 14 April 1978).
150 See, for details, McLoughlin, *op. cit.* in note 164, at p.102 *et seq.*
151 See further, *ibid.* p.106 *et seq.*
152 *Ibid.*
153 *Ibid.*, p.108.
154 For a description of the regulations, see Chapter IX, section 9.2. A list of the regulations is provided in McLoughlin, *op.cit.* in note 148, at pp.296−298.
155 EEC Council Directive No.70/220/EEC on the approximation of the laws of the Member States relating to measures to be taken against air pollution by gases from positive ignition engines of motor vehicles (OJ No.L.76, 6.4. 1970, p1) as adapted to technical progress by EEC Council Directive No. 74/290/EEC (OJ No.L159, 15. 6.74, p.61); EEC Council Directive No. 75/716/EEC on the approximation of the laws of the Member States relating to the sulphur content of certain liquid fuels (OJ No.L 307, 27.11.75, p.22).
156 See further McLoughlin, *op. cit.* in note 148, at p.74 *et seq.*
157 See text above, following note 84. For the latest draft articles on these subjects, see Part XII of the Informal Composite Negotiating Text, 15 July 1977 (ILM, Vol. 16, 1977 p.1108).
158 The EEC Commission presented proposals to Foreign Ministers early in May 1978 and in April 1978 IMCO's Maritime Safety Committee adopted an Anglo-French traffic scheme to keep laden tankers at least 30 miles from Ushant. This scheme, due to be introduced in January 1979, has been criticised by Trinity House, the U.K. lighthouse and pilotage authority, which has itself devised an alternative 200-mile shipping "Motorway" through the English Channel (*The Times*, 16−18 May 1978).

159 According to Resolution A.237 (VII) of the IMCO Assembly of 12 October 1971, the main objectives of the International Conference on Marine Pollution, 1973, were the achievement, by 1975 if possible but certainly by the end of the decade, of the complete elimination of the wilful and intentional pollution of the seas by oil and the minimisation of accidental spills.

Appendix

STATUS OF CONVENTIONS ON OIL POLLUTION AS AT 1 MAY 1978

Note: Regional conventions have been excluded from this table.

Key: 'X' means that the State has ratified or otherwise accepted the convention and that the convention is in force.

'O' means that the state has ratified or otherwise accepted the convention but that the convention has not yet attracted sufficient acceptances to bring it into force.

1.	OILPOL 1954-69	International Convention for the Prevention of Pollution of the Sea by Oil, 1954, as amended in 1969.
2.	Barrier Reef 1971	1971 (Great Barrier Reef) Amendments to 1 above.
3.	Tanks 1971	1971 (Tanks) Amendments to 1 above.
4.	HS 1958	Geneva Convention on the High Seas, 1958.
5.	CS 1958	Geneva Convention on the Continental Shelf, 1958.
6.	SOLAS 1960	International Convention for the Safety of Life at Sea, 1960.
7.	SOLAS 1974	International Convention for the Safety of Life at Sea, 1974.
8.	SOLAS PROT 1978	Protocol of 1978 to 7 above.
9.	Intervention 1969	International Convention relating to Intervention on the High Seas in cases of Oil Pollution Casualties, 1969.
10.	CLC 1969	International Convention on Civil Liability for Oil Pollution Damage, 1969.
11.	Fund 1971	International Convention on the Establishment of an International Fund for Compensation for Oil Pollution Damage, 1971.
12.	COLREG 1972	Convention of the International Regulations for Preventing Collisions at Sea, 1972.
13.	MARPOL 1973	International Convention for the Prevention of Pollution from Ships, 1973.
14.	MARPOL PROT 1978	Protocol of 1978 to 13 above.
15.	Civil Offshore 1976	Convention on Civil Liability for Oil Pollution Damage resulting from Exploration for and Exploitation of Seabed Mineral Resources, 1976.

STATES PARTIES	1 OILPOL 1954-69	2 BARRIER REEF 1971	3 TANKS 1971	4 HS 1958	5 CS 1958	6 SOLAS 1960	7 SOLAS 1974	8 SOLAS PROT 1978	9 INTERVENTION 1969	10 CLC 1969	11 FUND 1971	12 COLREG 1972	13 MARPOL 1973	14 MARPOL PROT 1978	15 CIVIL OFFSHORE 1976
Afghanistan				X											
Albania				X	X										
Algeria	X	O	O			X				X	O	X			
Argentina	X					X						X			
Australia	X			X	X	X									
Austria	X			X		X						X			
Bahamas	X		O			X			X	X	O	X			
Belgium	X			X		X			X	X		X			
Brazil						X				X		X			
Bulgaria	X			X	X	X						X			
Burma						X									
Byelorussian SSR				X	X										
Cambodia (Democratic Kampuchea)				X	X	X									
Canada	X	O	O			X	X					X			
Cape Verde								O				X			
Central African Empire				X											
Chile	X					X				X		X			
China, People's Republic of						X									
China, Republic of				X											
Colombia				X											
Costa Rica				X	X										
Cuba						X			X						
Cyprus				X	X										
Czechoslovakia				X	X	X						X			
Denmark	X			X	X	X	O		X	X	O	X			
Dominican Republic	X			X	X				X	X		X			
Ecuador						X			X	X		X			
Egypt	X					X									
Equatorial Guinea						X									
Fiji	X			X	X	X			X	X					
Finland	X	O	O	X	X	X			X			X			
France	X	O				X	X	O	X	X		X			
Gabon						X									
Gambia						X									
German Democratic Republic				X	X	X				X		X			
Germany, Federal Republic of	X			X		X			X	X	O	X			
Ghana	X					X						X			
Greece	X	O	O			X	X			X		X			
Guatemala				X	X										
Guinea						X									
Haiti				X	X	X									
Honduras						X									

The role of law . 293

STATES PARTIES	CONVENTIONS														
	1	2	3	4	5	6	7	8	9	10	11	12	13	14	15
	OILPOL 1954—69	BARRIER REEF 1971	TANKS 1971	HS 1958	CS 1958	SOLAS 1960	SOLAS 1974	SOLAS PROT 1978	INTERVENTION 1969	CLC 1969	FUND 1971	COLREG 1972	MARPOL 1973	MARPOL PROT 1978	CIVIL OFFSHORE 1976
Hungary				X		X						X			
Iceland	X					X						X			
India	X					X	O					X			
Indonesia				X		X									
Iran						X									
Ireland	X					X						X			
Israel	X			X	X	X						X			
Italy	X	O	O	X		X									
Ivory Coast	X		O			X				X					
Jamaica				X	X	X									
Japan	X			X		X			X	X	O	X			
Jordan	X	O	O											O	
Kenya	X			X	X	X								O	
Korea, Republic of						X						X			
Kuwait	X					X									
Lebanon	X	O	O			X			X	X					
Lesotho				X	X										
Liberia	X	O	O			X	O		X	X	O	X			
Libya	X					X									
Malagasy Republic (Madagascar)	X			X	X	X									
Malawi				X	X										
Malaysia				X	X	X									
Maldives						X									
Malta	X	O	O		X										
Mauritania						X									
Mauritius				X	X										
Mexico	X			X	X	X	O		X			X			
Monaco	X					X	O		X	X		X			
Mongolia				X											
Morocco	X					X			X	X		X			
Nauru						X									
Nepal				X											
Netherlands	X			X	X	X			X	X		X			
New Zealand	X	O		X	X			X	X		X				
Nicaragua						X									
Nigeria	X			X	X	X						X			
Norway	X	O	O	X	X	O		X	X	O	X				
Oman						X									
Pakistan						X						X			
Panama	X					X	O		X	X					
Papua New Guinea						X						X			
Paraguay						X									

STATES PARTIES	CONVENTIONS														
	1	2	3	4	5	6	7	8	9	10	11	12	13	14	15
	OILPOL 1954–69	BARRIER REEF 1971	TANKS 1971	HS 1958	CS 1958	SOLAS 1960	SOLAS 1974	SOLAS PROT 1978	INTERVENTION 1969	CLC 1969	FUND 1971	COLREG 1972	MARPOL 1973	MARPOL PROT 1978	CIVIL OFFSHORE 1976
Peru						X									
Philippines	X	O	O			X									
Poland	X			X	X	X				X	X		X		
Portugal	X			X	X	X					X				
Romania				X	X	X							X		
Saudi Arabia	X	O	O			X									
Senegal	X			X	X	X				X	X				
Seychelles						X									
Sierra Leone				X	X										
Singapore						X							X		
Somalia						X									
South Africa				X	X	X					X		X		
Spain	X			X	X	X				X	X		X		
Sri Lanka						X							X		
Surinam	X									X					
Swaziland				X	X										
Sweden	X	O	O		X	X				X	X	O	X		
Switzerland	X	O	O	X	X	X							X		
Syria	X	O	O			X				X	X	O	X		
Thailand				X	X										
Tonga				X	X	X	O						X		
Trinidad & Tobago				X	X	X									
Tunisia	X	O	O			X				X	X	O	X	O	
Turkey						X									
Uganda				X	X										
Ukranian S.S.R.				X	X		O								
United Kingdom	X	O	O	X	X	X	O			X	X	O	X		
United States	X			X	X	X				X			X		
Uruguay	X					X									
U.S.S.R.	X	O	O	X	X	X				X	X		X		
Upper Volta				X											
Venezuela	X			X	X	X									
Vietnam						X									
Yemen, People's Democratic Rep. (Sana'a)	X					X									
Yugoslavia	X		O	X	X	X				X	X	O	X		
Zaire						X							X		
Zambia						X									
TOTALS	59	20	21	56	54	97	12	0	31	34	12	53	3	0	0

Note *The Fund Convention (No. 11 in the table entered into force in October, 1978.*

14

Conclusion

J. Wardley-Smith

The reader was unfamiliar with many aspects of oil pollution will by now have come to the conclusion that oil pollution can never happen. Each chapter has discussed at some length the various precautions which must be taken to avoid accidental spills and reference has been made to the various codes of practice regulations and local bye-laws which have been drawn up by all the various authorities with the object of making oil pollution impossible. In addition chapter 13 described some of the laws and penalties which are applied to those who, for one reason or another, allow oil to be spilt. Surely all of these must combine to make the spilling of oil, on land or water, a very rare occurrence. On the other hand, casual observation of the news media will soon bring to light cases where oil tankers have run aground sometimes spilling oil, where collisions between tankers, or between tankers and other vessels have occurred and, usually on a more local scale, where road tankers have left the highway, overturned and caused pollution. A walk along a sandy sea-shore in almost any part of the world will soon bring to view lumps of congealed oil.

While this book was being written several major incidents have occurred. In the first a road tanker loaded with a highly flammable petroleum derivative ran off the road into a crowded holiday camp on the coast of Southern Spain, the resulting fire and explosion killed 200 people. In another incident, the largest marine oil spill to date, the *Amoco Cadiz,* carrying 220,000 tons of light crude oil, had a steering gear failure while rounding the coast of Brittany *en route* for Rotterdam via Lyme Bay, where it would have been lightened. There was some delay in the arrival and use of a salvage tug, the weather worsened and finally the vessel broke up completely losing all its oil onto the shore of Brittany. Thus, not only do accidents still occur, but as the quantities transported increase the size of each consignment, so the size of the problem which arises, should an accident occur, becomes even worse.

One of the most tragic tanker accidents of all time occurred early in January 1979 in Bantry Bay, Eire. A small fire started on the French tanker *Betelgeuse* while it was unloading its cargo of 120,000 tons of crude oil. Almost at once the half emptied tanker blew up killing all 43 crew members and 7 men working on the jetty. The fire burned for two days before the stern portion sank and it was extinguished. Fires and explosions whilst loading or unloading are fortunately very rare. In the past a member of VLCCs have had explosions during tank working at sea. This type of accident has diminished in frequency due to improved operating techniques and, in particular, the use of the inert gas systems to replace air in empty cargo tanks. At present (January 1979) the cause of the *Betelgeuse* disaster is unknown, but it may be significant that the ship was not fitted with the inert gas system.

There have also been two major tanker accidents polluting the coast of the United Kingdom. On 6 May 1978 the Greek tanker *Eleni V* carrying 17,000 tons of heavy fuel oil was cut in two by

Plate 26: *The* Betelgeuse *after the explosion in Bantry Bay.*

the French bulk oil carrier *Roseline*. The stern containing about three quarters of the oil was towed away and salvaged, but the remaining 5,000 tons was spilled and considerable coastal pollution resulted. The dispersants available were unsuited for the heavy oil, and the organisation of the whole salvage/clean up operation was not as effective as it might have been. In the other incident the *Christos Bitas* ran aground close to the coast of Wales on 12 October 1978. It was about 10 miles away from a normal course, the vessel proceeded towards the Irish coast before summoning assistance, some 3,000 tons of its cargo of 35,000 tons of Iranian crude were lost. Highly successful arrangements, aided by good weather enabled the remaining oil to be off loaded into other tankers and finally the hull was towed into the Atlantic 300 miles west of land and scuttled in deep water. Coastal pollution was largely prevented by the prompt use of dispersant spraying vessels which were well directed and could deal with the light crude oil.

These two spills indicated again that prompt, well organised oil salvage is of fundamental importance in reducing pollution. Proper deployment of spraying vessels will, if the oil is not of a heavy type, do much to reduce actual beach pollution.

Previous chapters in this book have discussed the movement of oil from the well to the final consumer of the product. Many and various examples have been given of the way in which oil, or the product, can be spilt or accidentally discharged. In all of these cases there has been one common factor, one reason given, which probably accounts for more than seventy per cent of all the incidents that occur. This is human error. It could be argued that when more sophisticated automatic, self checking and self calibrating equipment can be installed, then the position will improve. This is probably true, but not nearly to the extent which might be expected. For example, the almost universal use of radar, which displays the coastline and other vessels within range, should have made collisions in fog a thing of the past; unfortunately this is far from being the case. Two of the largest, most modern and best equipped tankers, the *Ven Pet* and *Ven Oil* collided in fog off the Cape of Good Hope early in 1978. In fact, when radar first became standard equipment on ships, a new sort of accident arose which was called 'radar induced collision.' Better training and improved equipment have reduced this type of accident but, as we have seen, they can still happen. Elaborate equipment requires better and more highly skilled maintenance. No longer can the ship's crew or the plant staff repair anything that fails and even duplication of equipment that might fail is no solution.

There is at least one example of a tanker fitted with two independent radar systems coming to a grief because both of them were unservicable! The solution to this problem lies in the human factor discussed in chapter 12. Human factor engineering, to use the American term, began during the Second World War, when the new sophisticated weapons were not producing the results their designers had expected and accidents to their operators, particularly to aircraft pilots in training, became far too frequent. There had been a failure to allow for the imperfections and slow responses of many human senses. These wartime problems were solved by bringing together two groups of specialists: those who knew about human capabilities and those who knew about machines. Anatomists, physiologists and experimental psychologists worked together with engineers and weapons specialists to make man plus machine an effective fighting weapon. The success of this teamwork has resulted in the extended application of ergonomics in the armed forces. This, too, has extended to industry; it will be clear from Chapter 12 that there are many instances in the vast oil industry where ergonomic principles have not been adopted.

Not only must the equipment be improved — the lay out of controls, the design of display panels, and even the colours of indicator lamps and knobs — but the whole of the process in question must be examined to see that the different warnings and signals given to the operator are those which his fallible senses can use to the best advantage, and this must be in conjunction with expert and thorough training.

Plate 27: *Control room of the BP tanker* British Respect

The bulk carriage of oil, and indeed of hazardous polluting substances in general, has highlighted the conflict between those who believe that the international waters of the oceans are free to all, where anyone may go where they like without let or hindrance, only risking their own ship and crew, and those who believe that in the modern world there is need for careful international control and routeing of all aspects of shipping whether it be tanker, passenger cargo or even pleasure vessels. Where shipping routes pass through territorial waters, then the State can draw up its own routeing or other arrangements and it has full power to see that they are enforced. Some such routes, as for example the English Channel, are considered to be international highways and thus the state laws do not apply.

The Inter-Governmental Maritime Consultative Organisation has done a great deal in drawing up recommendations, codes and specifications dealing not only with problems of oil pollution but with matters such as the saving of life at sea, provision of life boats and the training of ships crews. Regulations and instructions of this sort only come into force when a sufficient number of nations have signed the Convention to bring it into force and then it is only applied by those nations which have signed the convention and subsequently passed the necessary national legislation to make the proposed rules mandatory. States which have not signed the convention have no obligation to bring the changes into force. Commercial pressure can of course be applied, charterers may be forbidden by those states who have signed the Convention to charter vessels which do not comply, but even this fairly simple move can meet with difficulties and it is frequently the case that the state which is trying to obey the rules finds that its freight charges have gone up, with the result that imported goods cost more than would have otherwise been the case. The process for enforcing these international laws is also long and tedious. Any infringement is reported to the flag state of the vessel concerned which then has to take appropriate action in its courts. The proposal that the port state should have full jurisdiction within its territory which is being discussed at present at the Law of the Sea Conference should make enforcement much easier.

Undoubtedly the discussions at IMCO of the technical recommendations, made after long and careful study of the problem by the world's experts, are all moving in the right direction, towards the reduction of accidents, loss of life and pollution. The process is slow, but governments can be encouraged by local pressure groups such as the Advisory Committee on Oil Pollution of the Sea in the UK and the Nordic Union in Sweden. Commercial and national interests perhaps unfortunately play a great part in these discussions which should ideally be confined to technical matters.

The provision of segregated ballast tanks in oil tankers for example, would be one satisfactory way of virtually eliminating operational pollution. Tanker owners are generally in favour of this as it would mean a reduction in the carrying capacity of about 15 per cent with a consequent need for 15 per cent more voyages. Countries with ship building and ship repairing interests are also in favour of the proposal because of the value to their yards of the modification programme which would result. On the other hand, however, governments which have to import a large percentage of their oil are not in favour, as the 15 per cent reduction in carrying capacity would mean a corresponding increase in the price of the product. This is only one example, but in nearly every case good economic and quasi-technical arguments can be raised on each side; it can be seen that reaching a consensus can take a considerable time. As we have said already spills do occur, and what is more they will continue to occur until the next generation of equipment, machines and of course, operating personnel have been introduced everywhere. Steps must be taken therefore to mitigate the nuisance, danger and discomfort which can be caused by an oil spill. Some of these points have been touched upon in previous chapters and the major methods of dealing with spilled oil have been dealt with in a companion volume to this book.[1] What can be re-iterated here is the need for everyone concerned to have considered the possibility that a spill may occur — the chance may be remote, but it can still happen. Having made this assumption it must then be decided what action would have to be taken to deal with such a situation. In some hazardous fields this is a matter of routine. All oil refineries have a fire brigade and all of them have practised dealing with various possible incidents. Equipment is available to deal with at least the initial outbreak, and plans are made so that, should it become necessary, outside help can be called without delay. Regular training sessions are held, safety committees are set up to examine all of the processes and operations to see which of them could, if something went wrong, produce a hazard. The net result of all these activities is to reduce the chance of accidents happening and, when they do happen, to deal with them so quickly and so efficiently that the small incident, the small fire, does not escalate into the large incident, the holocaust.

An organisation such as has been described has obviously been set up when the risk, particularly to life, is very high. Nevertheless, the same principle must be adopted wherever oil or petroleum products are stored, transported, or manufactured, for it is essential that all the people who might be concerned with the incident know exactly what task they must undertake. The chain of command must be defined quite clearly.

Everyone must know what actions they must take, their instructions must be clear and unequivocal. A single command must be responsible for all operations both on land and sea so that all the separate actions can be co-ordinated and scarce resources of men and materials used to the best advantage to mitigate the pollution as a whole. Oil, whether spilt on land or water spreads rapidly. It is therefore essential that measures are taken as soon as possible to prevent a further spread contaminating a larger area. Valves must be closed, booms set up, absorbent materials applied, all as quickly as possible. This can only be done, when the people concerned know what they have to do, where they are going to do it and have had practice in doing it. In any scheme no matter how small training exercises must be held regularly. When personnel change, a new man cannot be expected to step into the shoes of someone who has left, and know all the background, he must be trained to work with the team. It is only by setting up local organisations wherever there is a risk, that the situation as a whole can be improved and those spills which do occur will have the least effect.

This book has briefly reviewed the harmful effects of oil and oil products when they are released on to land and water and also when they, or the products of their combustion, are discharged into the air. These effects can be seen to be harmful to many creatures living in the marine environment, and in some circumstances, harmful to man himself. This is probably the strongest reason why every effort should be made to avoid accidental discharges. The same arguments apply with even more force to the many other hazardous chemicals which are now being transported in ever increasing quantities and which are listed in the 1973 Marine Pollution Convention.

There now seems little doubt that the end of what might be called the age of oil is in sight: discoveries of new oil fields will in all probability extend the actual period, beyond some of the 'final' dates now being given, but everyone agrees that there is not an unlimited amount of oil in the world and that it will come to an end within a finite time. The major oil producing countries have already increased very considerably, the price of oil and prices are likely to increase still further as it becomes less plentiful. There are, therefore, two further reasons why every effort should be made not to spill oil. It is a wasting asset that should not be thrown away, it is a valuable product which should not be wasted. This is not the place to discuss what it should be used for; is it too valuable a resource to be used merely as a source of heat and fuel for motor cars? What can be said, however, is, that every care must be taken not to waste it by spilling it on land or water.

REFERENCES

1. For further details see J. Wardley-Smith (ed) *The Control of Oil Pollution* (Graham & Trotman, London, 1976).

Acknowledgements and sources

Grateful acknowledgement is made to the following organisations and individuals for permission to reproduce the illustrations included in this book.

CHAPTER ONE
Figure 1   G.A.B. King *Tanker Practice* (Stanford Maritime Ltd, 1976).
Plate 1   J. Wardley-Smith.
Plate 2   Dr. I. White, ITOPF Ltd.
Plate 3   J. Wardley-Smith.
Plate 4   Press Association.

CHAPTER TWO
Figure 2   A. Nelson-Smith.
Plate 5   A. Nelson-Smith.
Plate 6   A. Nelson-Smith.
Plate 7   Dr. J.M. Baker, Oil Pollution Research Unit, Orielton Field Centre, Pembroke.

CHAPTER THREE
Plate 8   Warren Spring Laboratory.
Plate 9   Warren Spring Laboratory.
Plate 10   S.R. Craxford.

CHAPTER FOUR
Figure 3   D. R. Blaikely
Figure 4   D. R. Blaikely
Figure 5   D.R. Blaikely
Figure 6   D.R. Blaikely

CHAPTER FIVE
Figure 7   W.O. Gray
Figure 8   W.O. Gray
Figure 9   W.O. Gray
Figure 10   W.O. Gray
Figure 11   W.O. Gray
Figure 12   W.O. Gray
Figure 13   W.O. Gray
Figure 14   *Exxon Marine* vol 18 no 3
Figure 15   *Exxon Marine* vol 18 no 3
Figure 16   W.O. Gray
Figure 17   Exxon
Figure 11   TNO, Institute for Mechanical Constructions, Delft
Figure 18   W.O. Gray
Figure 19   IMCO

CHAPTER SIX
Plate 12   British Petroleum Co. Ltd.
Plate 13   British Petroleum Co. Ltd.

CHAPTER SEVEN
Figure 21   R.S. Hawkins
Plate 14   J. Wardley-Smith
Plate 15   J. Wardley-Smith
Plate 16   J. Wardley-Smith
Plate 17   British Petroleum Co. Ltd.

CHAPTER EIGHT
Figure 22   Gulf Oil Company
Plate 18   Santa Fe International Corp.

CHAPTER NINE
Plate 19   British Petroleum Co. Ltd.
Plate 20   British Petroleum Co. Ltd.
Plate 21   J. Wardley-Smith

CHAPTER TEN
Plate 22   Esso Petroleum Co. Ltd.
Figure 23   H. Jagger
Figure 24   H. Jagger
Figure 25   H. Jagger
Plate 23   J. Wardley-Smith

CHAPTER ELEVEN
Figure 26   G.F. Oldham
Plate 24   Mass Transfer Ltd.
Plate 25   Mass Transfer Ltd.

CHAPTER TWELVE
Figure 27   W.T. Singleton
Figure 28   W.T. Singleton
Figure 29   W.T. Singleton

CHAPTER FOURTEEN
Plate 26   Press Association
Plate 27   British Petroleum Co. Ltd

Index

ADR, transport regulations 185, 187
API separator 216-217, 219
accidents: analysis of human 243; anxiety 238; deliberate 238; non-tankers 113; social theory 239
accident records 84-89
accidents to tankers, survey 9
acid rain in Scandinavia 53
acid smuts 47, 49
activated sludge method 224, 225, 227-228
activity analysis 224-245
advisory vessel traffic system (VTS) 109
aerosols in air 43
air pollution, definition 35
Air Quality Standard (USA) 44
Allegheny Mountains, long-distance pipeline 155
Ambient Air Quality Standard (USA) 44
American Petroleum Institute (API) specification for pipelines 155
Amoco Cadiz 9, 10, 95, 258, 271, 283
Anglo-Norwegian Agreement 258
animals, effect of oil on 20, 26
Arctic regions: effect of oil on vegetation in 19; special problems 19
Argo Merchant 95
Arizona Standard, oil spill 11
automobiles, pollution by 8

bacterial corrosion of pipelines 168
ballast, international law 270
ballast water — oil removal from 214, 231
ballasting tankers 115, 117-118, 121, 136
Baltic Convention 257
Bantry Bay 296
Barcelona Convention 263
Battersea power station 56
beach pollution 4, 29
Betelgeuse 296
bilge water 8, 9, 115, 117, 124-125
bio-accumulation of oil 12
biochemical oxygen demand (BOD) 224, 225
biological treatment of oily effluent 224, 225
birds, effect of oil on 7, 24, 26
blow-down 214, 215
blow-out: restoring control 73-75; source of pollution 67, 68
Bonn Agreement 275
bulk transport by road and rail 183
bulkheads 82, 90

bunds for storage tanks 206
bunkering 151, 153
bunker — transfer 116, 117, 125-126
butane 197

Canadian Petroleum Association 156
carbon dioxide as a pollutant 36, 37
carbon monoxide as a pollutant 37
carbon monoxide emitted per ton of fuel 37
carcinogens in oil 12, 13
cargo: discharging 136; loading 137-139; operations 135; pumps 136; transferring 128
cargo hose 144
catalytic devices for automobiles 45
chemical treatment of effluent 229
chimney height 51, 53
chocolate mousse 28
Civil Liability Convention, 1969 276
Civil Liability Convention, 1976 264
Civil Liability Convention and UK 266
classification of petroleum products 197
clean-up of oil spills, effect of 32
Clean Seas Guide 118
coalescers 221
coastal states, powers of control 274
Collision Avoidance Systems 112
Collision Regulations, 1972 271, 272, 273
communication, ship to shore 144
compartments, in tankers 90-91, 117
compensation 265-266
Compensation, Convention on, 1971 276
CONCAWE 156
conduct rules, offshore operations 255
containment, during oil transfer 142, 143, 151
Continental Shelf Act 1964 260
continental shelf, and law 255, 279
Control of Pollution Act, 1974 281
Convention on Compensation, 1971 276
Convention — Oil Pollution 1954 117
corrosion of pipelines 167, 175; bacterial corrosion 168
CRISTAL 278-279
crew, size and quality 84
crude oil tank washing 123-124
crude oil, toxic properties of 6

damage to pipelines 169, 178, 181
Decca, radio navigation system 112
decision theory 237

Department of Transportations Office of
    Pipeline Safety, US 156
Deposit of Poisonous Wastes Act, UK 210
desulphurisation of fuel oil 54
development of wells 61-67
diesel smoke 48
disc skimmer 218
Discharge of Oily Substance, 1973
    Convention 257
drilling of wells 61-67

EEC directive on waste oil 280
E&P activities 61
economic effects of pollution 35
effluents, re-use of 214
effluent, steelworks 231
effluent treatment 216; biological 224, 225;
    chemical 229
electrolysis of pipelines 168, 176
electro-osmosis 20, 22
Elizabeth Watts, 2
emulsified oil 231
environment, oil in 11
equipment on tankers, summary 149
equipment, testing of 150
error — human, reduction of 245
error reduction: personality and social
    aspects 246; procedures 246; skill and
    systems theories 248; system and
    operator 250
estuaries 29, 31; pollution of 31
Europe, pipeline spills in 13, 156
exhaust smoke 48
exploration and exploitation of continental
    shelf 255-279

filters: granular 225; percolating 224, 225;
    to remove oil 223
fish, effect of oil on 24, 26, 27
floating roof of storage tanks 201, 202, 203
flocculation 221
flotation: dissolved air 222; induced air 223;
    separation of oil 162, 222
flow-rate meters 170
flue gas, removal of sulphur from 56
fluidised combustion of sulphur-containing
    fuel 57
forest damage by SO_2 54
Fund Convention 1971 276, 278

galvanic corrosion of pipelines 167
gauges for storage tanks 203
Geneva Convention 1958 256
Gluckauf 2

hazard rating 23, 184
Hazchem code markings 185

Helsinki Convention 263
herbicide, oil as a 18
Home Office Code for Storage 200
hoses 137; inspection of 141; maintenance
    144
human error, as factor in spills 142-143,
    208; in pipeline operation 166, 174;
    statistical analysis 241; theories 234
human operators, overloading of 236
human performance theories 235-237
hydrocarbons as pollutants 40
hydrostatic pressure tests 170

IMCO 93, 95, 96, 97, 100, 108, 114, 123
ITIA 278
inert gas system 123; and safety 97-100
information theory 236
Institute of Petroleum Code of Practice
    for Storage Tanks 204, 261
International Chamber of Shipping (ICS)
    Guide 118, 271
International Convention 1954-71 26-30,
    254, 255, 269-270
International Convention 1969 274
International Convention 1973 257, 270
International Convention 1978
    Protocol 257, 270
International customary law 258-259
international law: definition of 254-255;
    land pollution 279-280
international rules, offshore operations 256,
    259-260, 274

Jakob Maersk 11
jetty: construction of 138; lighting 144;
    operation 143-147
Joint Utilities Location Information for
    Excavators (JULIE) 181

LPG 197, 198, 200, 203; storage 200
lagoons, treating effluent 228
Land Registry, UK 180
land spills 13, 15, 279
lay-barge 162
lead: as a pollutant 45; effect on health
    46
lead-free petrol 45, 46, 47
leak detection in pipelines 170-173
liability, industry schemes 278-279
Liability of Operator, 1976 Convention
    264-265
liability rules: international 263-266,
    276-277; municipal 266-267, 277-278
lightening of tankers 126
lighting of jetty 144
load-line 89-90, 93

load-on-top (LOT): monitoring 118; operating 121
loading of cargo: tankers 135-137; rail tank wagons 188; road tank wagons 188, 190
Loran, radio navigation system 112

maintenance of hoses 144
manoeuvring of tankers 101-108
mechanical failure, pipeline 166, 173
Mediterranean Convention 257
malodour 57, 58
mangroves, effect of oil on 29
Merchant Shipping Bill 1978 274
Merchant Shipping Act, 1894 273
Merchant Shipping Act, 1974 273, 278
Metula 11
Minerals Working Act 1971 261
monitoring LOT 118
mooring tankers 137
mud, oil based drilling 77
municipal law 255, 260; land pollution 280-282; offshore operations 260, 266, 272, 274

natural hazard, and pipeline failure 167, 174-175
natural seepages of oil 7
navigation 108
nitric oxide formation during combustion 41
nitrogen dioxide: concentrations in air 41, 43; WHO limits in air 42
Norwegian law: comparison with UK 262; decrees 1975, 1976 262; offshore pollution 267
OPOL 265-266
offshore operations: conduct rules 255; international rules 256, 259-260, 274
Offshore Pollution Liability Agreement 265-266
oil discharges — source of 213, 214
oil effects: on birds 24, 26; on fish 24, 26, 27; on land 18; on salt marsh 19; on shell fish 17, 27; on vegetation 18, 19
Oil in Navigation Waters Act, 1971 274
oil pollution law: enforcement 253, 269, 270; scope 254
oil pollution, world's 2
oil record book 130, 269, 273
oil spill: clean-up effects 32; removal 75, 76; prevention 208-211; regional co-operation 275
oil transfer: operations 125-128, 151; planning of 150

oily waste, selection of treatment method 230
Omega, radio navigation system 112
'One-call' system, 180
operation of jetty 143-147
Operation Discharge, 1973 Convention 256
operational pollution from wells 76-78, 117
operational speed of ships 103
operator limitations 32
operator skills 237, 248
Oregon Standard oil spill 11
organic oxidants in the air 43
ozone concentration in Los Angeles 43
ozone and oxidants 42
ozone, WHO limit in air 44

Paris Convention 1974 279
percolating filters, treatment of effluent by 224, 225
personality of operators 237-239
peroxyacyl nitrates (PAN) in air 43, 44
Petroleum and Submarine Pipeline Act 262
petroleum investment: hazard rating 198; classification of 197-198
Petroleum Production Regulations 1976 261
Petroleum Spirit (Consolidation) Act 1928 282
Petroleum Spirit (Conveyance by Road) Regulations 186-187
Petroleum Spirit Storage Regulations, UK 200, 201
photochemical oxidants in air 42
Pipelines: design factors 157; history 153-155; leak detection 171; leak prevention 171; legislation in UK 157; natural hazard 167, 174-176; operation 164-165; routeing in West Europe 158; route planning, 158; spills 13, 15; system malfunction 166, 174; third party damage 169, 178, 181; underwater 161-164; welding 160
plankton, effect of oil on 27
planning oil transfer operation 150
poisonous concentrations of carbon monoxide 37, 38
pollution: annual oil loss undersea production 8; effluent 8, 213-214; of beaches 4, 29; of estuaries 31; of the land 13, 15, 213, 279; of the land, laws on 279-282; non-tanker 113; operational from wells 76-78
Pollution Convention, 1973 132
potable water, effect of oil on 20

pressure testing, pipelines 170
Prevention of Oil Pollution Act 1971 254, 255, 260, 272, 274, 277
prevention of oil pollution law 253-254, 272
primary treatment of effluents 216
product tankers 84
production of oil in world 2
propane 197
Protocol-Oil Pollution, 1978 133
public opinion and mass media 234
pulled method, in underwater pipelining 162
pumps: cargo 136; shipping 136

radar 109
road tanker, design features 186
radio navigation systems 109, 112
rail tanker wagon design features 186
rain water, oily 214
re-use of effluents 214
regional co-operation in oil spills 275
road, bulk transport by 183
road tank wagons, spills while unloading 191
rail tanker, wagon size 184
reel barge 164
relief well 74
rivers, pollution by oil 22, 23, 24
road vehicles, size limits 184

safety precautions: while loading road/rail tankers 192; while unloading road/rail tankers 193-194; while vehicle loading 192-193; while vehicle unloading 193-194
safety regulations 93
safety regulations, IMCO 83, 95, 96, 97, 100, 108, 114, 123
salt marsh, effect of oil on 19
sea, effects of oil on 25
seabirds, effect of oil on 7
sea valves, leakage 136
seaweed, effect of oil on 28
secondary, biological treatment 224
seepage: natural 7; under sea 7, 8
segregation of oily effluents 215
separators: API 216-217, 219; circular 219-220; parallel plate, 220; tilted plate, 220; corrugated plate 220; plate 220; three chamber 220
shellfish, effect of oil on 17, 27
ship handling simulators 101
ship to ship transfer 126-128
ship — shore transfer 141, 142, 143, 151
shore vegetation, effect of oil on 29
skill, of operators 240

Sliktrak computer model 76
slop tanks 120, 137
smog 36
smoke and smuts 47, 48
smoke from diesel engines 48
smoke — UK Motor Regulations 48
social theory, and accidents 239
SOLAS convention 271, 272
spillage: accidental 271; deliberate 269; from non-tankers 113; human error 208; from pipeline 156; prevention 173, 208-211; road and rail tankers 186, 192-195; system malfunction in pipeline 166-174; on tankers 126, 129
steel pipelines 155
steelworks, effluent 231
Stockholm Declaration 268
storage, hazards in 199
storage tank 201; types 200
stripping pumps 136
Suez Canal 4, 105
sulphur burnt in UK/annum 49
sulphur dioxide in towns 50
sulphur, oxides of 49; UK regulations on air/pollution by 51
systems theory 239-40

tainting of fish by oil 20, 27
tank cleaning 97, 117-118, 119-120, 123-124
tank, location, foundation, pipework 203-206
tank wagons spills while unloading 190
tank washing — crude oil 123-124
tankers: accident survey 9; ballasting 115, 117-118; discharges from 8, 9; mooring 137; and the law 267-271; lightening 126; spillages 126, 129
tanks: horizontal cylindrical 201; moulded rectangular 201; slop 120; types of storage 201; vertical cylinder 201, 204
tar balls 4
tertiary treatment, & effluent 229
testing of equipment 150
thrusters, small tankers 105
Titanic 93
Torrey canyon 11, 29, 271, 275
TOVALOP 278
toxic properties of crude oil 6, 7
toxicity of oil to winkles 17, 23
traffic control 108
transfer bunker 116, 117, 125-126
transfer of oil 125-128; from ship to ship 126-128; vessel to shore 135
transport regulations 185
transportation costs & tanker capacity 83

UNCLOS III 283
United Kingdom Electricity Authorities, chimney smoke 53
UK regulations on sulphur emissions 51
United Nations Environment Programme (UNEP) 280, 283
ULCCs, accidental pollution 81
Urquiola 11

VLCCs: accidental pollution from 81; construction & efficiency 81, 82, 102, 105, 106
valves: inspection of 141; sea, leakage 136; on storage tanks 202, 211
vegetation: effects of oil on 18, 19, 29; in arctic regions 19; shore, effect of oil on 29

vehicle emission regulations in USA 38, 40
vehicle loading & unloading precautions 10, 192, 193-194
Ven Oil 10
Ven Pet 10
vents for tanks 202
Voluntary Code, 1976 132
Water Act, 1945 282
Water Act, 1973 281
water, fresh, effect of oil on 20
Water Resources Act, 1963 281
wells: blow-out 67, 68, 73-75; safety equipment 69-73
WHO, air pollution & health 44, 49
Woods Hole Oceanographic Institute 25
world oil movements by sea 79
work place design 246

MIX
Papier aus verantwortungsvollen Quellen
Paper from responsible sources
FSC® C105338

If you have any concerns about our products,
you can contact us on
ProductSafety@springernature.com

In case Publisher is established outside the EU,
the EU authorized representative is:
**Springer Nature Customer Service Center GmbH
Europaplatz 3, 69115 Heidelberg, Germany**

Printed by Libri Plureos GmbH
in Hamburg, Germany